Studies in Logic
Logic and Cognitive Systems
Volume 24

Studies in Diagrammatology and Diagram Praxis

Volume 14
New Approaches to Classes and Concepts
Klaus Robering, editor

Volume 15
Logic, Navya-Nyāya and Applications. Homage to Bimal Krishna Matilal
Mihir K. Chakraborti, Benedikt Löwe, Madhabendra Nath Mitra and Sundar Sarukkai, eds.

Volume 16
Foundations of the Formal Sciences VI. Probabilistic Reasoning and Reasoning with Probabilities.
Benedikt Löwe, Eric Pacuit and Jan-Willem Romejin, eds.

Volume 17
Reasoning in Simple Type Theory. Festschrift in Honour of Peter B. Andrews on His 70th Birthday.
Christoph Benzmüller, Chad E. Brown and Jörg Siekmann, eds.

Volume 18
Classification Theory for Abstract Elementary Classes
Saharon Shelah

Volume 19
The Foundations of Mathematics
Kenneth Kunen

Volume 20
Classification Theory for Abstract Elementary Classes, Volume 2
Saharon Shelah

Volume 21
The Many Sides of Logic
Walter Carnielli, Marcelo E. Coniglio, Itala M. Loffredo D'Ottaviano, eds.

Volume 22
The Axiom of Choice
John L. Bell

Volume 23
The Logic of Fiction
John Woods, with a Foreword by Nicholas Griffin

Volume 23
Studies in Diagrammatology and Diagram Praxis
Olga Pombo and Alexander Gerner

Studies in Logic Series Editor
Dov Gabbay   dov.gabbay@kcl.

Studies in Diagrammatology and Diagram Praxis

Olga Pombo
Alexander Gerner

© Individual author and College Publications 2010.
All rights reserved.

ISBN 978-1-84890-007-3

College Publications
Scientific Director: Dov Gabbay
Managing Director: Jane Spurr
Department of Computer Science
King's College London, Strand, London WC2R 2LS, UK

http://www.collegepublications.co.uk

Original cover design by orchid creative   www.orchidcreative.c
Printed by Lightning Source, Milton Keynes, UK

All rights reserved. No part of this publication may be reproduced, stored in a retrieval system or transmitted in any form, or by any means, electronic, mechanical, photocopying, recording or otherwise without prior permission, in writing, from the publisher.

Index

Preface i

I. Fundamental Issues on Diagramatology

Olga Pombo: *Operativity and Representativy of the Sign in Leibniz.* 1

Sybille Krämer: *'Epistemology of the Line'. Reflections on the Diagrammatical Mind* 13

Jan Wöpking: *Space, Structure, and Similarity. On Representationalist Theories of Diagrams* 39

Frederik Stjernfelt: *The Extension of the Peircean Diagram Category. Charting the Implications of a Diagrammatical Revolution in Semiotic* 57

Ahti-Veikko Pietarinen: *Is Non-visual Diagrammatic Logic Possible?* 73

II. Visual Diagram Praxis

Augusto J. Franco Oliveira: *Picture-proofs in Mathematics: a Chapter in James Robert Brown's Philosophy of Mathematics, with Examples* 85

Ángel Nepomuceno-Fernández: *Concept Script and Diagrams: Two Ways of Using Images in Scientific Reasoning* 95

J. R. Croca: *Diagrammatic Thinking in Physics* 117

Matthias Bauer: *Diagrammatology, Scenographic Media, and the Display Function of Art* 125

III. On Maps

Valeria Giardino: *The World in Maps: Looking for Treasures, Neurons, and Soldiers* — **145**

Alexander Gerner: *Notes on Diagrams and Maps* — **167**

Christoph Ernst: *"From Images to Diagrams". Diagrammatic Reasoning in Gilles Deleuze's Film Philosophy and its Relevance for General Media-Theory* — **191**

Catarina Pombo Nabais: *Deleuze's Rhizome-Thought* — **211**

Preface

This volume puts together most of the papers given at the interdisciplinary workshop on **Diagrammatology and Diagram Praxis** held at the University of Lisbon, 23-24th March 2009. The workshop was organized by the Research Project Image in Science and Art of the Center for Philosophy of Science of the University of Lisbon (CFCUL), namely by its task number two whose aim is to analyze the place of image and diagrammatic thinking in three different epistemological and semiotic programs (Leibniz, Frege and Peirce).

Unfortunately, the volume does not include the papers delivered by Irene Mittelbergs (HumTec RWTH Aachen), Paulo Urbano (University of Lisbon) and Miguel Cardoso & Santiago Ortiz (bestiario.org), Tiago Lança (architect), and Nuno Nabais (University of Lisbon). But we were able to incorporate additional papers of four invited speakers that unluckily were not able to be present in the Lisbon: Sybille Krämer (Freien Universität Berlin), Christoph Ernst (University of Erlangen-Nürnberg), Valeria Giordano (Institut Jean Nicod, Paris) and Anghel Nepumoceno (University of Sevilha). The articles are organized in three main sections. In the first, **Fundamental Issues of Diagrammatology**, we open with a paper by Olga Pombo untitled *Operativity and Representativity of the Sign in Leibniz*. The paper offers an historical account of the roots of diagrammatic thinking in the cognitive conception of language and in the semiotics which Leibniz worked out in the XVII century. Olga Pombo claims that it is necessary to go back to Leibniz in order to understand the requirement of both operativity and representativity of the sign as an attempt to combine the heuristic virtualities of written characters as stable and manipulative devices with its capacity of making visible, or as Leibniz says, of "painting" (pingi) the signified ideas in all its presence and complexity of relations.

Somehow in the same line - but now located in the pos-iconic turn - the paper of Sybille Krämer, *Epistemology of the Line. Reflections on the Diagrammatical Mind*, faces the need of recognizing the constitutive contributions of iconicity to our faculty of cognition. The stressed idea is that we are today confronted with a multiplicity of hybrid phenomena which combine language and imagery, digital and the analogical. On that basis, the author stresses the need of enlarging the concept of diagrammatic in order to embrace "all forms of intentionally created markings, notations, charts, schemes and maps". But, if so, it is necessary to recognize the biological and anthropological nature of graphic activity as a spatial-visual-tactile form of thinkin.

In the line of representationalist approaches, Jan Wöpking, *Space, Structure, and Similarity. On Representationalist Theories of Diagrams*, proposes to think the cognitive power of diagrams on the basis of two independent claims: the first concerns what he calls nomic similarity, a kind of structural correspondence (in Leibnizian terms, we would say "expressive") between the diagram and what it represents; the second relays on the two-dimensional graphical dimension of diagrams, their spatial schematic structure. Two claims and key distinctions which are used by Jan Wöpking in order to analyze and classify the several definitions of diagram and to revise some of Peirce's ideas on diagrams.

Frederik Stjernfelt, who gave the opening lecture in the 2009 Lisbon workshop, offers a paper titled *The Extension of the Peircean Diagram Category. Charting the Implications of a Diagrammatical Revolution in Semiotics*. First this revolution launched by Peirces' diagrammatology, as described by Stjernfelt, will be ontological – because it reemphasizes the connection of semiotics to mathematical and logical idealities as well as empirical universals. Stjernfelt argues that the uses of prototypical signs in logics, algebra, pictures, linguistic grammar and linguistic semantics are diagrams, thus opposing the common sense notion of diagrams as a printed or drawn appearing on a support as concrete sign tokens, that in his view merely give access to the type of diagram. The other "revolutionary" aspect of this view is that diagrams are fundamentally cognitive phenomena, involved in giving access to generalization, idealization, abstraction and thus providing direct perceptual access to an "ideal type". Stjernfelt builds his argument on Peirce with cross-reference to Husserl, as for Stjernfelt this cognitive notion of the diagram is providing direct perception-like access to ideal structures, that in Husserlian terms is described as "categorical intuition".

Ahti-Veikko Pietarinen, *Is Non-visual Diagrammatic Logic Possible?* develops the very interesting possibility of making a propositional logic based on sounds. For that, he goes back to Peirce and to his large conception of diagrams in order to show that diagrams should not be restricted to visual forms of representation. The question is thus formulated in order to question the extreme possibility of constructing a mental yet diagrammatic logic without any visual or written support, no symbols, no marks, no language.

The second part, **Visual Diagram Praxis**, includes four case studies on diagram practices, namely in mathematics, logics, quantum-physics and scenographic media.

Augusto J. Franco de Oliveira, *Picture-proofs in Mathematics: a Chapter in James Robert Brown's Philosophy of Mathematics, with Examples*, questions the role played by pictures, figures or diagrams in mathematics, identifies some of their purposes as visual aids, focus their disturbing or even highly misleading effects and passes in review some of the conflicting positions concerning this issue. From one side, Euclides' belief in geometric pictures, from the other Bolzano's efforts to avoid the intuitive properties of continuous functions, from one side the geometric intuition of curves underlying Newton and Leibniz calculus, from the other Dedekind, Cantor and Weierstrass endeavors in order to ground infinitesimal calculus in purely analytical terms, from one side Lagrange "paranoid" fear of figures and Dieudonné severe refusal of any geometric intuition, from the other side Brown's proposal of pictures as possible rigorous evidence. As Franco Oliveira shows, "with the development of computers, computer generated graphs (for instance, in fractal geometry), experimental mathematics and computer generated proofs in mathematics the discussion has been revived". However, as he also states, the relevant question persists: Is visual thinking merely a psychological aid, facilitating grasp of what is gathered by other means, or does it also have epistemological functions, as a means of discovery, understanding, and even proof?

Ángel Nepomuceno-Fernández, in a paper untitled *Concept Script and Diagrams: Two Ways of Using Images in Scientific Reasoning*, compares the conceptual writing proposed by Frege with Peirce's diagrams stressing the complementary nature of those two forms of picturing logical structures.

The paper by J. R. Croca's, *Diagrammatic Thinking in Physics*, emphasizes the need of using diagrammatic thinking in fundamental quantum physics. Face to the huge, say infinite, number of solutions for differential equations describing the physical situation, diagrammatic thinking appears as able to provide good reasons for choosing the right solutions.

Finally, Matthias Bauer, *Diagrammatology, Scenographic Media, and the Display Function of Art*, argues that diagrammatic reasoning is not restricetd to Logics and that art, in its fundamental display function, is itself a diagrammatic procedure. The papper offers then an interpretation of Rembrandt's famous painting "The Night Watch", of Peter Greenaway's "Nightwatching-project", and of Oliver Stone's feature film "JFK" as three examples of artistic works with clear scenographic or diagrammatic nature.

The third part, **On Maps**, opens with an article by Valeria Giardino, *The World in Maps: Looking for Treasures, Neurons, and Soldiers*. Enlarging the concept of map from geographical maps to new cases of maps such as cortical maps in neuroscience and graphs, Valeria Giardino reflects upon the different components that must be considered in a map, its specific spatiality, its unavoidable incompleteness. The map appear then as a procedure which necessarily lies and, at the same time, are able to provide a more or less precise description of what it is.

Alexander Gerner, *Notes on Diagrams and Maps* is an exploratory journey that hinges on the very question of what a map is. Gerner draws on the work of Stjernfelt, Perkins, Krämer and, mainly, Deleuze and Peirce. It is his argument that these authors' work open up seminal ways of interpreting diagrammatic fixtures common to different notions of the map - as a diagrammatic spatial knowledge tool or as an ontological category preceding any representational semiotic operations. The diagrammatic operational and the ontological map reading, though evidently at cross-purposes, should not ultimately be seen as mutually exclusive. The journey into the map territory in the second part of the paper reflects on contemporary map artists (Beltrán, Barrio, Hirschhorn & Steinweg) and on their artefacts. Art seems to surge as the place where a map can be accepted as the "Map of itself" (Terry Atkinson & John Baldwin 1967). However, that is not to say that this notion of self- indexicality of the proper map is absent in scientific maps. It is simply not visible, focused on or accepted as such. The "all knowing map" that Perkins (2008) describes as "scientific", is seen here as a contradiction towards the notion of map as a fundamental diagrammatic or ontological orientation tool, that necessarily has to leave out certain "knowledge" to constitute orientation. With Deleuze, even the "chasm" between different operational cultures of map use (Perkins 2008) becomes questionable.

Christoph Ernst, *"From Images to Diagrams" – Diagrammatic Reasoning in Gilles Deleuze's Film Philosophy and its Relevance for General Media-Theory*, stresses the deep similarity of Peirce's concept of diagrammatic reasoning and Deleuze's concept of the thinking-image. In general, Christoph Ernst's paper aims to "illustrate the relevance of Peirce's semio-pragmatic concept of diagrammatic reasoning for the further clarification of the category of the thinking-image in Deleuze's film-philosophy". According to Deleuze and in the line of his critique of rationalist understanding of human reasoning, thinking-images can no longer be described in terms of "figurative" images. Within the medium of film, thinking may now externalize its own potentialities, it may now produce the visualization of its own knowledge, it can be observed "in progress", happening on screen. We understand why Deleuze's film-philosophy supposes a strong reference to Peirce's diagrammatic reasoning. Two main questions may then be formulated: do Peirce's ideas on diagrammatic reasoning help clarifying Deleuze's film-philosophy? How does Deleuze's film-philosophy, or media theory in general, benefits from making use of Peirce's ideas?

Finally, Catarina Nabais in an article untitled *Deleuze's Rhizome-Thought*, presents Deleuze's specific non-representational account of the rhizome. Instead of an arborescent, hierarchical and static model of thought, the "rhizome" appears as an endlessly bifurcating system pointing to a nomadic movement of thought, a dynamic and heterogeneous way of thinking in process.

Stressing the Deleuzin thesis of the creative, experimental character of thought, Catarina Nabais shows in which way Deleuze cannot accept Peirce's reduction of diagram to icons, indexes and symbols. She points for two main reasons: first because Deleuze aims to emphasize, not the signifier-signified relations but the territoriality-deterritorialization relations; second because Deleuze wants to deny the representative nature of diagrams and to highlight their capacity for constructing "a real that is yet to come". We understand why for Deleuze,"diagram is a creative device. Not a graphic representation of relations but a creative agent of reality". Able to threaten the arborescent western culture, the concept of rhizome appears thus as a whole new diagram of thought, not moved by a good will of truth, but a constructivist performance or experience of becoming, that is, a model which proposes new political ways of thinking by multiple, subterranean and innovative proliferations.

The paper from Catarina Nabais functions thus as a proposal for an "open end" to this book and to its attempt of dealing with many unanswered question on Diagrammatology and Diagram praxis.

Olga Pombo and Alexander Gerner

I. Fundamental Issues on Diagramatology

Operativity and Representativity of the Sign in Leibniz

by
Olga Pombo

In opposition to, or in the line of, Philosophy cannot grow except inside a tradition. This is precisely the issue with the contemporary research on diagrammatic thinking. We all agree that it is essential to go back to Peirce or Husserl not only to recognize important roots, to honour significant predecessors, but to recover their insights, to make use of their hypothesis, to recuperate the conceptual instruments they have been able to put forward.

What I propose here is to go back even more in time, to go back to Leibniz. Not to claim for Leibniz as a predecessor of cognitive semantics or of present-day diagrammatic research, but to recall what Leibniz thought out in this respect, to bring into play the conceptual devices he put up, to bear in mind the distinctions he was able to constitute.

1. My first claim is that the historical importance of Leibniz philosophy of language is mostly due to his cognitive conception of language, that is, his recognition of the constitutive symbolic nature of human thought. As he says in the celebrated *Dialogus de connexione inter res et verba* (1677):

> "I will never be able to know, to discover, to prove without using words or without the presence in my mind of other signs" (GP 7.191)

This cognitive conception of language was mainly formulated in the scope of the debate against Descartes and of their different mathematical experiences. Descartes tends to start from Geometry, in which figural representation is a merely auxiliary, imaginative support for reasoning. Leibniz stars from algebra where operations are fully symbolically performed, where symbols constitute the reasoning process itself, completely replacing the supposed direct experience of mathematical objects.

In fact, the Cartesian principle of evidence entails a merely instrumental and communicative conception of language. For Descartes language has above all a communicative function. The most language may be asked to do is to operate as a mnemonic support for the recall of the long chain of reasons. Leibniz refuses the intuitionism of Descartes. For Leibniz, rigour cannot be dependent neither on subjectively based certainties, nor on the confidence in the intuitive infallibility of natural light. As Leibniz says in a *Letter to Gallois (1677),*

> *[The Cartesian methodological rules]"give surely beautiful decrees but not the guideline for developing those decrees"* (GP7.21)

According to Leibniz, rigour should be achieved by the use of a symbolic system which would render visible the more abstract ideas and would constitute a material support for thought and reasoning, a *"filum Ariadnes"* (GP7.22)[1], a *"filum palpabile"* (GP7.57, 59, 125), a *filum cogitandi* (C; 420), a *filum meditandi* (GP7: 14), a *filum mechanico* (C: 351), that is, a symbolic criteria, a manipulatory device. Something which Mathematics has already developed and which, as Leibniz used to say needs nothing but paper and ink. Let us look for what Leibniz writes in the well known *Preface à la Science Générale (1677)*

> *"Now the reason why the art of demonstrating has been until now found only in mathematics (...) is this: Mathematics carries its own test with it. For when I am presented with a false theorem, I do not need to examine or even to know the demonstration, since I shall discover its falsity a posteriori by means of an easy experiment that is, by a calculation, costing no more than paper and ink.*
>
> *The tests or experiments made in mathematics to guard against mistakes in reasoning (...) are not made on a thing itself, but on the characters which we have substituted in place of the thing.*
>
> *Take for example a numerical calculation: if 1677 times 365 are 612,105, we should hardly ever have reached this result if it were necessary to make 365 piles of 1677 pebbles each and then finally to count them all in order to know whether the aforementioned number is found.*
> *This test is performed only on paper, and consequently on the characters which represent the thing, and not on the thing itself". (C 154)*

---

[1] «La veritable méthode nous doit fournir un *Filum Ariadnes,* c'est à dire, un certain moyen sensible et grossier, qui conduise l'esprit, comme sont les lignes tracées en geometrie et les formes des operations qu'on prescrit aux apprentifs en Arithmetique», (GP7.22)

Symbolism is for Leibniz the proper, the necessary and essential means of human reason. Leibniz even goes a step further by claiming that only symbolic signs allow us to operate with ideal significations which can only be established by the sign and that only sign enable us to think. The example of the polygon of thousand faces presented by Leibniz in his celebrated *Meditationes de Cognitione, Veritate et Ideis* (1684*)* is eloquent:

> *"When I think of a polygon of thousand faces, I do not always consider what is a face, an equality or the number thousand but I use this words (whose meaning is present on my spirit only in a very confuse and imperfect way) in order that they (the words) take the place of the ideas which I have of them (...) I call this knowledge as blind or symbolic. We make use of it in algebra and arithmetic and in almost all domains"* (GP 4. 423).

Again, Leibniz mathematical experience lies on the basis of this important theory of blind thought (*cogitatio caeca*). In infinitesimal calculus, sign does not evoke but fully substitutes notions which human mind cannot otherwise completely reach. So, I believe, we are here face to two main correlated leibnizian theses which deserve to be recalled by our current research on diagrammatic thinking: a) the non merely communicative but cognitive conception of language; b) the theory of blind though.

2. My second claim is that it is necessary to be aware of the specificity of Leibniz semiology, namely the prospective, heuristic potentialities which Leibniz attributes to symbolic systems. In a *Letter to the German mathematician Walter von Tschirnhaus* dated from May 1678, Leibniz writes:

> *"No one should fear that the contemplation of characters will lead us away from the things themselves; on the contrary, it leads us into the interior of things. For we often have confused notions today because the characters we use are badly arranged; but then, with the aid of characters, we will easily have the most distinct notions, for we will have at hand a mechanical thread of meditation, as it were, with whose aid we can very easily resolve any idea whatever into those of which it is composed. In fact, if the character expressing any concept is considered attentively, the simpler concepts into which it is resolvable will at once come to mind. Since*

> *the analysis of concepts thus corresponds exactly to the analysis of a character, we need merely to see the characters in order to have adequate notions brought to our mind freely and without effort. We can hope for no greater aid than this in the perfection of the mind"* (GM 4. 461).

We find here some of the main thesis of the Leibniz semiology:

1. Characters do not disturb the knowledge of reality.
2. On the contrary, characters constitute a powerful means for the development of knowledge.
3. Characters not only reflect but also promote human knowledge allowing to go from confuse to clear, distinct notions.
4. This happens because of the written, graphic (iconic) nature of characters.
5. Characters are mostly valuable because they can be contemplated, because they can be spatially arranged, because they can be handled, manipulated, used as mechanical threads.
6. Characters could then become a mechanical thread of meditation.
7. But for that to be possible, it is necessary that characters may express concepts.
8. With such expressive characters, then the analysis of characters will correspond exactly to the analysis of concepts.
9. If this happens, then to seeing the characters would be the easy way of getting adequate knowledge.
10. The result should be of the greatest importance for the perfection of human mind.

Leibniz is always very clear concerning the recognition of the qualities and potentialities coming from the spatial nature of characters. And he extends that recognition, not only to the characters of the new artificial philosophical language to be constructed, but also to natural languages, namely to writing. The most illuminative passages in this respect appear in the *Nouveaux Essais*. While he recognizes the critical arguments presented by Lock's Philalethe against the difficulties, insufficiencies and disturbing effects put forward by natural languages[2], Leibniz position consists

---

[2] In fact, the XVII century discusses deeply the role which language performs in the process of knowledge: Does language help to promote knowledge? Or, on the contrary, is language a disturbing factor for the acquisition of knowledge? Two great positions can be signalized. A critical position, which emphasizes language's insufficiencies and ambiguities: Bacon, Locke, Descartes, Arnauld, Melabranche and, in general, all those who look for the construction of

invariably in stressing the virtues of language and most of all the merits of writing for a more rigorous and regulated utilization of the various dimensions of signification. As he writes:

> "*Mais pour revenir à vos quatre defauts de la denomination, je vous dirai, Monsieur, qu'on peut remedier à tous, surtout depuis que l'écriture est inventée*" (Nouveaux Essais, III, IX, § 9).

Leibniz project is thus double. He aims: 1) to overcome the difficulties of natural languages by the exploration of their capacities in terms of definition and writing, 2) to construct a new language, more exactly, a writing system or *Characteristica Universalis* of which Mathematics would be just an example. As Leibniz says in a *Letter to Gallois (December, 1678):*

> *"But to make it easier and, so to speak, more tangible, I intend to make use of the Characteristic, of which I have spoken with you on occasion, and of which Algebra and Mathematics are merely examples. This Characteristic consists of a certain writing or language (since he who has the one can have the other) which perfectly corresponds to the relations of our thoughts. This characteristic would be completely different from any that has been envisaged until now, since the most important thing has been overlooked, which is that the characters of this writing have to serve for discovery and for judgement, like in algebra and in arithmetic"* (A II 1.669)

Again, it is clear that Leibniz praises the properties of writing. Only its two-dimensional spatiality makes possible a more clear and differentiated expression of relations.

We are here face to the enormous and very much ambitious project of a *Characteristica Universalis*. All along his life, Leibniz will be fully committed with this program which he claimed to be very much innovative[3]. In fact, Leibniz does not avoid criticism towards his predecessors in terms of the construction of new philosophical language, the line that goes from Lull to Kircher and that which goes from the English

---

new artificial languages, and a positive position which, although recognizing some limits and imperfection of human languages, nevertheless stresses its constitutive character. From my point of view, just two names in modern times: Thomas Hobbes and Leibniz.

[3] Something which is more significant since, as well known, Leibniz was in general very much interested and attentive to all the achievements of his predecessors.

pasigraphers to Dalgarno and Wilkins.[4] However, according to Leibniz, none of his predecessors have managed to guarantee the heuristic virtualities that symbolism may offer. And that by two main reasons: the insufficient analysis of thoughts which, in Leibniz opinion, underlies the set of *predicamenta* or *suma genera* in the basis of which their systems were constructed[5] and the arbitrary nature of the characters they have established[6]. That is to say, on the contrary of what is could appear, Leibniz is not a full formalist. He does not fall into the illusion that the automatic functioning of a set of operational rules can permit the development of science. On the contrary, Leibniz claims that the new symbolic system should be semantically opened to the reality that it must permit to say. According to Leibniz, beyond responding to a logical objective of faithfully and rigorously express thought and its articulations, characters should open the road to the progress of knowledge. And that heuristic capacity can only be achieved if the system of characters to be constructed could be directly open to the reality they are supposed to say. Above and beyond operativity and functional capacity of the signs within the formal system, characters should also be "natural" that is representative of the world which is to be known. Leibniz heuristics involves a semantic exigency.

3. My third claim intends precisely to call attention to the representativity of sign which Leibniz wanted to conquer for the characters of the philosophical language to be constructed.

With this requirement of the representativity of the sign, Leibniz looks for the isomorphism between characteristic signs and the reality they signify. But how could it be done? How is it possible for a sign to represent reality? And which reality should be represented? Its sensory, particular traits as they appear to our perception? Its essence? The composition of its elements? The inner relation of its parts? Given the extreme difficulty of such a task, it is easy to understand the heterogeneity of this leibnizian

---

[4] Leibniz knows well all their works. He praised above all Wilkins's project, *An Essay towards a Real Character and a Philosophical Language, with an Alphabetical Dictionary*, published in 1668, with the support of Royal Society. As Leibniz writes: «J'ai consideré avec attention le grand ouvrage du Caractere reel et Langage Philosophique de Mons. Wilkins; je trouve qu'il y a mis une infinité de belles choses, et nous n'avons jamais eu une Table des predicateurs plus accomplie» (*Letter to Burnet, 24th August 1697*, GP 3: 216).
[5] See, for instance, GP 3.216.
[6] See for instance, C. 177 ff.

theory of the representativity of the sign. I will try to briefly summarise the several models which I believe Leibniz has pursued[7].

The first model is the extreme proposal of a figurative representativity as the pictorial representation of the sensible, imagetic traits of the signified reality. As Leibniz says in the *Nouveaux Essais*:

> "*Et on pourrait introduire un caractère universel fort populaire et meilleur que le leur, si on employait de petites figures à la place des mots, qui représentassent les choses visibles par leurs traits, et les invisibles par des visibles qui les accompagnent, y joignant de certaines marques additionnelles, convenables pour faire entendre les flexions et les particules*" *(GP 5. 379).*

A procedure which Leibniz believes to be applicable to visible and also to invisible realities.

> "*Those (realities) which cannot be figured [pingi], like the intelligible, should however be represented by some hieroglyphic method at sometime uniform and philosophical. This can be done if we do not pursue particular similarities, as the painters, the mystic and the Chinese do, but if we follow the idea of the very thing*" *(GM 5: 216).*

In second and weaker model, representativity is conceived, not as the figuration of the particular traits of the signified reality as they appear to our perception, but as its essence, that is, as the direct representation of the essential basis of those particularities. A good example is the project which Leibniz calls a *Characteristica Realis (GP7.12-13)* in which the character would represent the unifying principle or "key" (*clavis*) of the multiple properties which compose the signified idea.

> "*The name of each thing will be the key of all we should say, think and reason about that thing (...). The name we will give the gold will be the key of all we can humanly know about gold, that is, by reason, and according to such an order that, by the examination of that name, we can discover which experiences should be rationally realized with that name*" *(GP 7.13)*

---

[7] For a developed presentation of the various models of representativity in Leibniz and which we present here in a condensed, abridged way, see our Pombo (1987: 174-190)

There is, finally a third model, which I propose to call expressive representativity, aims to discover, neither the figuration of sensory particularities of the signified reality (figurative representativity), nor the essential basis of those particularities (essentialist representativity) but rather the analogical and structural reproduction of the network of relations constituting the idea and its articulations.

Even if rigorously speaking all representativity is in Leibniz expressive (since expression is for Leibniz the supreme instance of any kind of relation), it is now Geometry which provides the privilege model for the relation of expression. In fact, Geometry is characterized by the immediate establishment of an isomorphism between the idea represented, as an ideal structure, and its schematic figuration. But this isomorphism does not imply similarity, as the imitation of the original.[8] That which expresses does not have to be similar to the thing expressed provided that some kind of analogy can be discerned between them. The reaction of expression can even accept dissimilarity. As Leibniz says in *Essais de Théodicée:*

> *"The same circle can be represented by an ellipse, a parabola, an hyperbola, another circle and even a strait line or a point. Nothing seems so different, neither so dissimilar, than those pictures; however, there is an exact relation of each point to each point" (GP 6. 327)*

Now, it is within this expressive model that Leibniz points to a purely diagrammatic form of representativity. The most interesting indication is present in the fragment *Essais d'Analyse Grammaticale* (1683/4) where Leibniz considers to be licit to connect the components of characters by different lines since, in this way, he says, it would be possible "to see" all those components in a simultaneous way. It is worth wile to give the word to Leibniz:

> *"It would be licit to connect by different lines the parts of the character since, in this way, it would be possible to see the in paper in a simultaneous form while the sound speech vanishes and, thus, the first sound cannot refer the posterior unless the second contains something which corresponds to what it was in the first" (C. 285)*

It should be noted that the recognition of the value of writing in contrast to speech which Leibniz presents here, it is not limited to the

---

[8] In fact, this isomorphism implies a search, not for similarities, but rather, as Leibniz puts it, for *"un rapport constant et reglé entre ce qui se peut dire de l'une et de l'autre"(GP 2.112)*

stressing of writing's advantages in terms of fixing and registering or as a support for the failures of attention or of reasoning. What Leibniz here emphasizes is the two-dimensionality of writing, the possibility of simultaneous grasp of the multiple relations which characters (and the ideas they signify) establish among them. That is, diagrammatic symbolism, to which the passages cited above tend, has the merit of permitting the simultaneous apprehension of the relations among the entities represented.

In another passage, Leibniz introduces a slight distinction within expressive representativity. As he writes in the celebrated *Dialogus de connexione inter res et verba*, of August 1677:

> *"Even if characters are arbitrary, their use and connection have nonetheless something which is not arbitrary, that is to say, a certain proportion between characters and things, and, at the same time, between the diverse characters which express the relations of things among themselves"* (GP 7.192)

Leibniz is looking here for the possibility of displacing the ideal of representativity from the character (without however completely abandoning it) on to the plane of syntactic relations. The meaning of a character comes now to depend either on its integration in a formal structure – its *usus et connexio* – or on the correspondence (*proportio*) between this structure and the structure of the real which it aims to represent. Now, it is the very form of relations between characters that is seen analogical with things in the form of their relations.

Ultimately, there is – I will further argue – another last possibility which Leibniz explores in the fragment *Characteristica Universalis* of 10th August 1679:

> *"The more exact are the characters, that is, the more they represent the relation of things, the bigger is their utility and if characters are able to exhibit all the relations of the things, as do the arithmetic characters that I use, then there will be nothing in the things which cannot be deduced from the characters"* (GM 5. 141, our emphasis).

Representativity is again, not between structures (syntactic and natural) but between elements, between the individual (containing relations as its predicates) and the character as a sign which, through its own flexion, would "exibit" the multiple relations of the thing. It is now again each character, and not the relations between characters, that is to express the relations which individual entities contain as their predicates.

Then, within the expressive model of representativity, Leibniz pursues three approaches. In the first, the structure is purely formal; in the second, the structure becomes a source of meaning because, in itself, it imitates the real conceived as a structure; in the third the structure is absorbed by the sign and it is through the sign that it becomes expressive. If, in the first tendency (clearly diagrammatic) the semantical level tends to be entirely reduced to the syntactic, and in the second tendency (strictly structural) priority is given to analogy between the formal structures of language and the structures of the real, the third and last tendency (from the figurative and the essentialist models) recovers the requirement of the semanticity of the sign itself, that is, of a space that should be symbolically differentiated and, as such, indicative of the signified reality, open to the world which is to be known.

As I have tried to argue, it is precisely this indicative power of the sign, this openness of the symbolic system to the signified reality that Leibniz wants to safeguard at all costs. That is why he has pursued so many models, has attempted so many different ways of, besides operativity, gaining such a representativity.

This is a thesis which Leibniz recuperates and pulls through a long tradition coming from Plato's *Cratylus*, the speculations on Adamic language all along the Middle Ages, Renaissance and modern times.

A thesis which will be advanced by Peirce's conception of the iconic nature of diagrams and of the revelatory, heuristic power resulting from the relational analogy which lives in their heart.

* * *

We know that Leibniz never achieved any new language system nor have succeeded to establish a system of characters with the openness he wanted them to be endowed. He just left projects, drafts, sketches, numerous fragments, *specimina*, *échantillons*. However, his unfinished work must be regarded, not so much as a root of recent research in the area, but as a source for clarification of the difficulties as well as of virtualities involved, as a font for the formulation of powerful hypothesis and of delicate distinctions.

References

Leibniz, *Gottfried Wilhelm Leibniz Samtliche Schrifften und Briefe*. Akademie der Wissenschaften zu Berlin. Reihe I - VI. Darmstadt: Reichl (1923 segs). **A**

Leibniz, *Opuscules et fragments inédits de Leibniz. Extraits des manuscrits de la Bibliothèque royale de Hannover par Louis Couturat*. Paris: Alcan, 1903. **C**

Leibniz, *G.W. Leibniz. Mathematische Schriften, Hrsg. v. Carl Immanuel Gerhardt*. 1-7. Hildesheim: Olms, 1962. **GM**

Leibniz, *Die philosophischen Schriften von Gottfried Wilhelm Leibniz. Hrsg v. Carl Immanuel Gerhardt*. 1-7. Hildesheim: Olms, 1960. **GP**

Pombo, O. (1987), *Leibniz and the Problem of a Universal Language*, Münster: Nodus PubliKationen.

Epistemology of the Line'
Reflections on the Diagrammatical Mind

by
Sybille Krämer

Thesis, Context and Assumptions

Diagrammatic inscriptions, among which we include graphic artefacts ranging from notations to diagrams to maps, are media that provide a point of linkage between thinking and intuiting, between the 'noetic' and the 'aisthetic'. By means of this interstitial graphic world, the universal becomes intuitable to the senses and the conceptual becomes embodied: the difference between the perceptible and the intelligible is thus at the same time bridged – and constituted.[1] This is our thesis.

This thesis is embedded in a context which we shall briefly sketch here. Today, images are recognized as a legitimized object of research in epistemology and philosophy of science. They are considered not merely a means to illustrate and popularize knowledge but rather a genuine component of the discovery, analysis and justification of scientific knowledge.[2] However, the assumption that icons create evidence did not begin with our contemporary computer visualizations. Rather, the sciences have always been dependent on the 'eye of the mind', which gives the register of the senses access to the invisible, the abstract, the hypothetical and the imaginary.

The epistemic rehabilitation of iconicity takes place in the wake of talk of an 'iconic turn'.[3] This 'iconic turn' articulates a call to correct the claim to the absolutism of the linguistic made by the 'linguistic turn', as well as to bring to the fore the constitutive contributions of iconicity to our faculty of cognition and to the existence of epistemic objects. The upward

---

[1] Peirce explicitly assigned 'diagrammatic reasoning' this intermediary function between intuiting and thinking, as Stjernfelt (2000 and 2007) has shown. Other dimensions of this synthesis can, however, be detected in the work of other philosophers, as this essay and Krämer (2009) demonstrate in the cases of Plato, Kant, Peirce, and Wittgenstein. Stekeler-Weithofer (2008) has recently restored geometry as a diagram-based structural model to the centre of mathematics, thereby rehabilitating Kant's synthetic a priori.
[2] As exemplified by Lynch (1998); Daston and Galison (2007); Bredekamp et al (2003); Bredekamp and Schneider (2006); Earnshaw and Wiseman (1992); Heintz and Huber (2001); Heßler (2006); Tufte (1997).
[3] 'Iconic turn', see Boehm (1994, 13); 'pictorial turn', see Mitchell (1992).

revaluation of iconicity thus draws on a conceptual binary: the differentiation between language and picture.[4] We are well acquainted with this differentiation, for which a wealth of related terms exists: discursive and iconic, telling and showing, representing and presenting, arbitrariness and resemblance, digital and analog, etc. Yet we are nonetheless confronted with a multiplicity of epistemically relevant phenomena that are not to be simply relegated to the side of language or the side of the image, which rather conjoin qualities of both the linguistic and the iconistic. This is the point of departure for the reflections to follow in this paper: there is a significant group of depictions that joins together traits of both language and imagery, albeit in varying proportions: Think of notation as language made visible; unutterable formal languages; music spatialized in the form of the score; diagrams that synthesize drawing with notation; or maps combining the digital and the analog. The attribute shared by all these image-language hybrids is the graphism of the line. This graphism dwells on the far side of imagery but this side of language. The remarks to follow shall centre on a subsection of it that is here called 'diagrammatics'. It is our goal to work out the epistemological significance of diagrammatics.

The knowledge-related functions of diagrams are not a new topic: publications shedding light on various facets of diagrams – primarily historically, but at times also systematically – have become abundant.[5] To date, however, a theory of diagrammatics remains absent.[6] It is precisely this missing theory to which we seek to contribute. Our 'diagrammatic perspective' is here connected with a historic-systematic re-framing that also concerns the extension of the concept 'diagrammatics'. We presume that formalism, understood as an operative notation that uses the two-dimensionality of surfaces to depict non-visual matter by means of visual configurations and to operate syntactically with this matter, is also to be considered a modality of the diagrammatic.[7] We thus include as diagrammatic not only diagrams in a narrow sense but all forms of intentionally created markings, notations, charts, schemes and maps. If this broadening of the concept 'diagrammatics' makes sense, the diagrammatic is a cultural technique[8], which – in its historic implementations – forms an indispensable basis for virtually all activities of human cognizance. Two

---

[4] Mitchell (1994, 5) pointed out, however, that there are no "'purely' visual or verbal arts" in the realm of the arts. See also Halawa (2008, 27ff.).
[5] As exemplified by Anderson and Meyer (2001); Bogen (2005); Bogen and Thürlemann (2003); Châtelet (2000); Gehring and Keutner (1992); Gormans (2000); Jamnik (2001); Lüthy and Smets (2009); Schmidt-Burckhardt (2009); Siegel (2009); Stjernfelt (2007).
[6] Lüthy and Smets (2009, 439); Mersch (2006, 103).
[7] Cf. Krämer (2003); Krämer (2005).
[8] On the concept of 'cultural techniques', see Krämer and Bredekamp (2003).

assumptions come into play here: (i) First, the anthropological supposition that the graphism of marks constitutes a defining feature of the human species. As a spatial-visual-tactile organizing form of thinking, arising from the coordination of eye, hand and mind, graphism is by no means secondary in significance to the cognitive role of language. (ii) Second, the supposition that not only sciences but even philosophy is twinned with diagrammatic structures. The diagrammatic features of the philosophical concept of reason remain largely neglected; reason, then, has yet to be reconstructed diagrammatologically. These assumptions form the backdrop before which we shall now investigate our thesis that diagrammatic artefacts act as a hinge between thinking and intuiting, while at the same time constituting the very differentiation between them.

We shall begin by elucidating two characteristic uses of diagrammatic operations in philosophy, Plato's Simile of the Divided Line and Descartes' coordinate geometry, as well as the figurative notation of Descartes' mathesis universalis.

Plato's Simile of the Divided Line

In The Republic (Politeia, 509 d – 511 e), Plato develops what is known as the 'Simile of the Divided Line', in which Socrates orders the ontological structure of the world according to an original-image relationship, and organizes this sub-categorization by degrees of knowability: one is to draw a line and divide it into two unequal subsections, such that the smaller subsection depicts the visible and the larger subsection the intelligible. These two subsections should then each be re-divided in the same proportion as the original division. A four-part division of realms of being is thus created, which at the same time embodies a series of levels of cognitive knowledge – that is, the activities of the soul – with a progressively increasing degree of theoretical clarity.

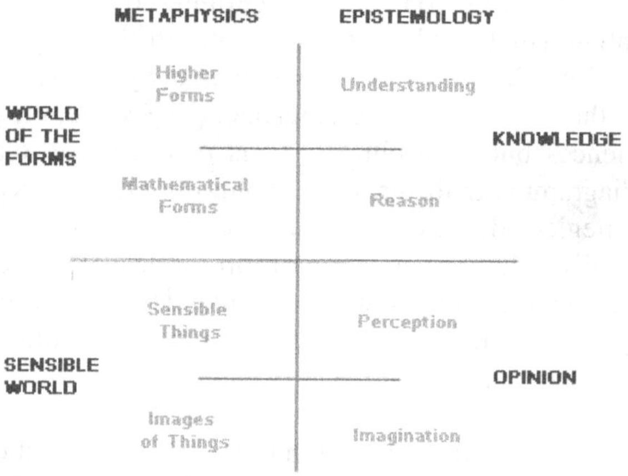

	METAPHYSICS	EPISTEMOLOGY	
WORLD OF THE FORMS	Higher Forms	Understanding	KNOWLEDGE
	Mathematical Forms	Reason	
SENSIBLE WORLD	Sensible Things	Perception	OPINION
	Images of Things	Imagination	

(http://www.molloy.edu/sophia/plato/divline_russo.gif)
Fig. 1 Depiction of the Simile of the Divided Line

Within the region of the visible, the lowermost realm is comprised of mirror-images, shadows and reflections in water, which correspond to the epistemic state eikasia, that is, conjecture. The next section of the visible encompasses the originals of these copies, material objects such as inanimate objects, plants and animals. The cognitive activity corresponding to these is pistis, faith or belief. Together, these two levels form the domain of 'doxa', that is, opinion. In the third subsection, which opens up the realm of the intelligible and thus of the 'episteme', reside general concepts and mathematical objects. The form of knowledge here is dianoia, that is, thinking or understanding. The fourth section, in turn, is concerned with Forms as true being understood through noesis, the act of pure reason (that is, intellection), which for Plato constitutes the highest level of understanding.

We are unable here to undertake any analysis of the Simile of the Divided Line in terms of the multiplicity of its interpretations and unanswered questions.[9] Rather, we wish to offer a summary of those aspects of the simile that make it a key moment for our understanding of diagrammatics. First, however, we must take a more precise look at the third level, where Plato locates mathematical objects, to him the protoform of scientific objects. As Plato characterizes this form of cognizance,

---

[9] Literature on the Simile of the Divided Line: Brentlinger (1963); Fogelin (1971); Nicholas (2007); Notopoulos (1936); Stocks (1911); Yang (2005).

mathematicians use visible objects as images representing invisible ideas: While mathematical speech and proofs refer to perceptible figures such as particular circles or numbers, they deal not with these concrete figures, but rather with the general concepts 'circle' or 'number', which are themselves not visible, but rather purely intelligible.[10] The form of cognizance of dianoia treats the visible as the imaging for the senses of something that is purely intelligible. It is thus the distinctive feature of mathematical knowledge to depend indispensably on the sensory representation of its theoretical objects and at the same time always to remain conscious of the difference between the intuitive and the purely intelligible. We want to stress four diagrammatologically instructive aspects.

(1) Iconicity as ontological principle: The capacity for making images – and with it visuality –forms the essence of Plato's ontology, understood as a doctrine of what is real: even the highest level of being – the Forms – are introduced as originals, thus as templates for pictorial copies. Degrees of reality are held up against the measuring stick of the original-copy relationship. Correspondingly, Plato first introduced to philosophy the term theoria, which originally meant 'Viewing of a festive performance'[11]. In counterpoint to Plato's perennially invoked hostility to images, to which philosophy's hostility to images was able to casually attach itself, it must be noted that iconicity is the inner principle of organization of Platonic ontology and epistemology.

(2) Differentiation between the visible and the intelligible; bridging of this differentiation: Plato differentiates categorically between the perceptible and the intelligible, and thereby introduces a differentiation that was for 2000 years to form the lifeblood of philosophy. The Simile of the Divided Line inaugurates this differentiation. At the same time, however, it identifies an area of epistemic activity – dianoia, characteristic of mathematics and sciences – in which this difference is intentionally bridged in that sensory objects are recognized as images of the non-sensory. Aristotle himself designated these Platonistic objects, on which mathematical action focuses but does not linger, as 'intermediary' or 'intermediate'[12]. 'Dianoia',

---

[10] Plato (1993, book vi, 510e): "They treat their models and diagrams as likenesses, when these things have likenesses themselves, in fact (that is, shadows and reflections in water); but they're actually trying to see squares and so on in themselves, which only thought can see."

[11] The original meaning of 'theoros' was the envoy of the polis sent to participate in festivals of the gods and oracles: see König (1998, 1128).

[12] Aristotle, *Metaphysics* I, 987b 14-17: "Further, besides sensible things and Forms there are the objects of mathematics, which occupy an intermediate position, differing from sensible

the activity of scientific reasoning, is – to express it in a modern way – to be understood as something conveyed by symbolic media. Even mathematicians need sensory representation in order to understand their noetic objects.[13] Decisive here is the fact that the sensory objects deployed in thinking are neither a final product nor an end-stage, but rather are passed through en route to that which is not perceptible but only intelligible. The bridging function of the third level as an 'interstitial world' consists in its enabling of this movement of thought.

(3) Spatiality of thinking: Plato views cognizance as an activity characterized by an implicit spatiality. Thinking is directional and this direction can be characterized as an ascent, a rise through ascending levels. Thus is the Allegory of the Cave in the next book of the Politeia (Republic), a continuation of the Simile of the Divided Line by other means: It visualizes not abstract lines but rather the concrete situation of a cave, and knowledge is here imagined as an ascent out of the cave and into daylight.[14] This inherent spatiality corresponds to the methodical matter that Plato, in the act of applying his own approach, deploys configurations of lines to visualize intellectual matters.[15] For Plato himself, the diagrammatic scene becomes the medium of insight. The diagram functions as an instrument of making evident the structure of ontology and epistemology.

(4) History of the reception of the Simile of the Divided Line: While Plato's text has been handed down to us without any illustration, virtually all interpreters have supplemented the text with concrete diagrams. In scarcely any other case in all philosophy do diagrams appear as frequently in texts as in the secondary literature treating the Simile of the Divided Line.

---

things in being eternal and unchangeable, from Forms in that there are many alike, while the Form itself is in each case unique."

[13] On this interpretation, see Krämer (1991, 53ff.).

[14] Plato, *Republic,* vol. vii, 514 -519.

[15] Also instructive here is the use of geometric diagrams: Menon 82b-84c and 86e-87a, Theaitet 147c-148d, Politikos 266b.

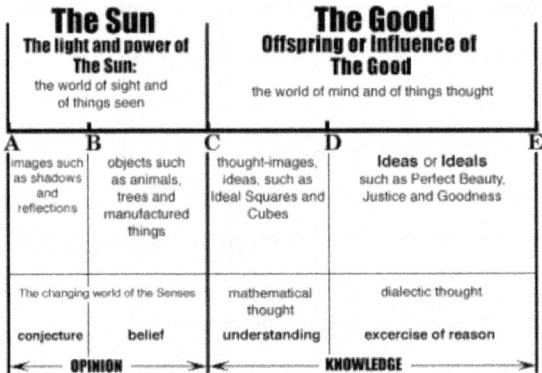

2a) (http://media.photobucket.com/image/Platon%20analogy%20divided%20line/terrymockler/Plato-Divided-Line.gif)

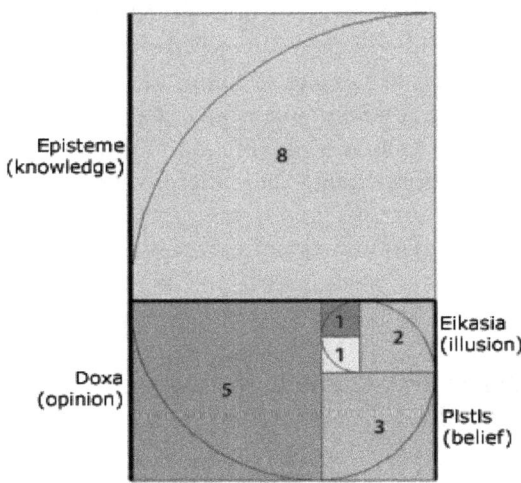

2b) (http://www.inctr.org/publications/images/2004_v04_n04_s01b.gif)

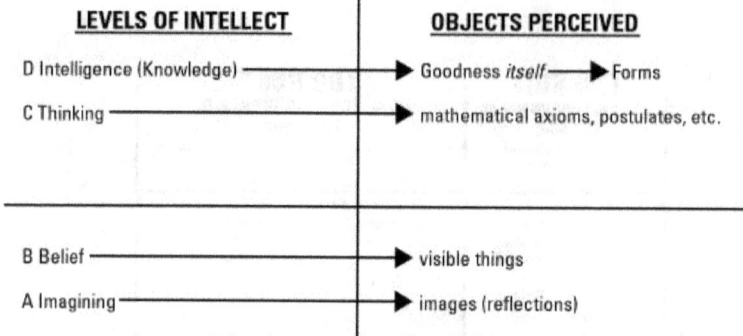

2c) (http://media.wiley.com/lit_note_images/154/1.jpg)
Fig. 2 Examples of Diagrams of the Simile of the Divided Line

Two features of these secondary visualizations are particularly striking: (i) The line that is to be subdivided is drawn in a variety of ways: horizontally or vertically, with the apportionment of the bottom subsections beginning on the left or the right. (ii) Equally multifarious and divergent are the interpretations of the Simile of the Divided Line.[16] It would be interesting to examine the relationship between the variations in interpretations and the differences in corresponding visualizations of the Simile of the Divided Line. We cannot undertake such an investigation here. We would like, however, to venture a conjecture to which this history of reception leads us: a line is not simply a line. Diagrammatic depictions turn on the localization and direction of the line on the two-dimensional surface, as this sets the course for reading and interpretation. At the same time, however, it is clear that diagrams do not interpret themselves.[17] A diagram is always bound up with multiple interpretations. And inversely, the same thought can be visualized in a great variety of ways, even in the case of the simplest line diagram.

Descartes: Coordinate System, Analytical Geometry, and mathesis universalis

We shall now jump over to the philosophy and mathematics of the Early Modern Era. Here, Descartes (1596-1650) provides us with the second key moment for the diagrammatic. Since the discovery of

---

[16] A compilation of various diagrams is found in Ueding (1992, 29ff.).
[17] Using a wealth of historical examples, Lüthy and Smets (2009) illustrate that diagrams are under-determined.

incommensurability[18] in ancient Greece, arithmetic and geometry, the countable and the measurable, had been treated as domains of mathematical objects to be vigilantly differentiated.[19] Descartes now prompts a minor mathematical revolution in that by virtue of a two-dimensional coordinate system[20] he rejoins the geometric figure to the algebraic formula. It is not the use of coordinates as 'reference lines' per se that is so consequential but rather the use Descartes makes of the coordinates in that through these lines he makes geometry and algebra jointly depictable and mutually translatable. He thereby initiates 'analytic geometry'.[21] Descartes published 'La Géomètrie'[22] as one of the three appendices[23] to his treatise on method, the 'Discours de la méthode'[24], in order to use particular scientific disciplines as examples demonstrating the usefulness of his general method. There is thus a genuine link between Cartesian 'coordinate geometry' and Cartesian epistemology; our supposition is that the linear form of diagrammatic inscriptions plays a decisive role therein.

As was done with Plato, with Descartes we must limit ourselves to a roughly outlined reconstruction and cannot proceed historico-philosophically.

(1) Coordinate geometry. Aided by the introduction of a coordinate-like reference system –which, however, is in Descartes not yet right-angled – number pairs can be assigned to points and can with clarity be numerically localized in their position on a surface. Geometric construction and problem solving can thus be transcribed in algebraic form and carried out as arithmetic problems. Descartes moreover replaced the ancient existence criterion that a geometrical figure be constructible with a compass and ruler with its algebraic calculability: curves that can be algebraically transcribed are genuine parts of scientific geometry; the others remain outside the science of mathematics.

But the mathematical effects of analytic geometry are of far less interest to us here than its diagrammatic imprint. We start here from the

---

[18] On this discovery, see Fritz (1965).
[19] More precisely, see Krämer (1991, 32-43).
[20] The diagram of the coordinate system and its geographic and mathematical history reach back to antiquity. Rottmann (2008) has recently reconstructed the traces of this history. Nicole Oresme (ca. 1323-1382), Pierre Fermat (1601-1665) and later Isaac Newton (1643-1727) are likewise central to modern coordinate geometry.
[21] Krämer (1989).
[22] Descartes (1902).
[23] Descartes (1902): La Dioptrique, Les Météores, La Géométrie.
[24] Descartes (1902, 1-77).

depiction of the now-familiar right-angled Cartesian coordinate system, named after Descartes although he made use of neither right angles nor the negative abscissa (x-axis) and ordinate (y-axis) axes. Lines are shaped into a configuration in the form of intersecting axes. The axes are numbered in ascending order. Numbers generally have no place in space. The line segment, however, allows for their simultaneous plotting and thus also placement. The axes have arrows, are thus directional, and subdivide the mathematical plane into four clearly defined and organized quadrants.

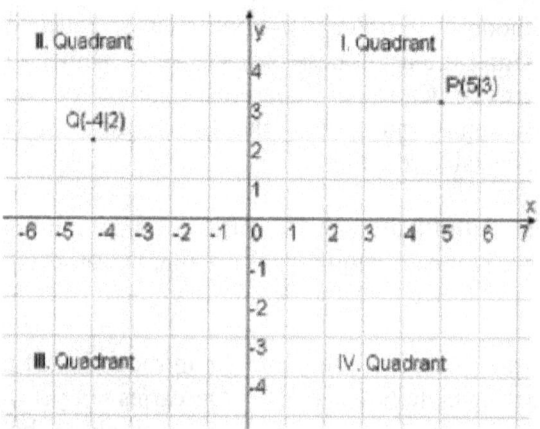

(http://de.wikipedia.org/wiki/Kartesisches_Koordinatensystem)
Fig. 3 The Right-Angled 'Cartesian Coordinate System', Named After Descartes

A fundamental diagrammatic insight becomes apparent here: the interaction of line and surface creates diagrammatic inscription's 'own space'. The arrangement of the axes enables a mathematical space to emerge from an empty surface. This space contains well-defined and delimited regions, the quadrants, in which the positions defined by the number pairs can be either occupied or unoccupied. Thus does a metamorphosis take place: through the inscription of coordinates, a sensorial apparent, real writing surface is transformed into a two-dimensional, mathematical, ideal plane. The points marked onto the writing surface have a perceptible extension, which is necessary to make possible the points' numerical identification and localization through coordinates. But on the ideal mathematical plane these points are without extension. Thus, we could say that the points realize a 'double existence' in coordinate space. They fulfil a double function as both a material trace of a marking (punctum, Lat. 'puncture') and an ideal construct of cognition. The coordinate system not only transforms an empirical surface into an abstract space but also creates a form of spatiality in which materiality and ideality, the intuitive and the intelligible, become intertwined. Hand, eye, and mind can work together on this diagrammatically produced surface and thereby first bring to life what we here shall call the 'eye of the mind'. In this sense, the coordinate system is a key to understanding diagrams in general.

(2) The transformation of geometry into a language. Coordinate geometry is a mechanism of translation: it makes two-dimensional geometric figures and linear algebraic equations mutually transferable. The term 'translation' has been chosen quite deliberately here: Descartes changes the very 'nature' of geometry in that he makes geometry 'compatible' with arithmetic. By virtue of the mediating position of the coordinate system, planarity turns into an identifying feature of analytic geometry, as it is based on the reduction of a three-dimensional geometry of solids to a two-dimensional plane geometry of line segments.[25] Descartes thereby turns geometry into a visual language, a transformation he is able to achieve by dispensing with the 'principle of homogeneity'.[26]

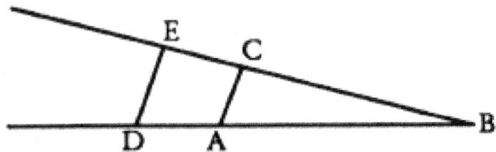

Fig. 4 Geometric Construction of Multiplication

"For example, let AB be taken as unity, and let it be required to multiply BD by BC. I have only to join the points A and C, and draw DE parallel to CA; then BE is the product of BD and BC." (The Geometry of René Descartes, trans. from French and Latin by David E. Smith and Marcia L. Latham, New York: Dover Publication, 1954, 5.)

The principle that two intersecting line segments determine a plane and two intersecting planes determine a solid had held since Greek antiquity. Descartes breaks from this principle: just as a number is created by the

---

[25] This is already proclaimed in the first sentence of La Géométrie: "Any problem in geometry can easily be reduced to such terms that a knowledge of the lengths of certain straight lines is sufficient for its construction."
[26] More precisely: Krämer (1989).

multiplication of numbers, the multiplication of line segments likewise results in a line. Descartes thus discards the 'figurative significance' of lines as structural elements of geometric figures and treats line segments as uninterpreted basic characters in calculations. Geometry becomes line calculation and Descartes' 'La Géomètrie' marks the birth of a language-centred understanding of the mathematical. Herein lies a key to the surge in quantification that occurred in the Early Modern Era: mathematics is no longer a realm of timeless, ideal objects, but rather advances to a type of universal and operative visual language, capable of expressing all that is quantifiable.

(3) Mathesis universalis: The constituting of scientific objects through mathematical language. It is in fact not until the posthumously published early form[27] of his treatise on method, the 'Regulae ad directionem ingenii'[28], that Descartes formulates a novel science that he names mathesis universalis, for which he sees forerunners in both ancient problem-solving analysis and the new symbolic algebra developed by Viète.[29] The mathesis universalis is a science of all quantifiable objects. Its particularity is that it works with an artificial system of notation in the form of a two-dimensional graphism of extensional figures, a medium through which anything that is to be made an object of this science must be depictable. This figurative 'language of the eye' is operative: it not only represents its objects but also makes them manipulable and thus generates and constitutes them, as the unity of objects arises from the unity of the operation.[30] This is demonstrated by Descartes' newly developed concept of a 'universal quantity', introduced as a reference object of his ideographic notation. As a result of the division between geometry and arithmetic, 'quantity' had always been subdivided into 'magnitudo', the geometric realm of the measurable, and 'multitudo', the realm of the arithmetically countable; there was no concept of a 'universal quantity'. But Descartes created as the reference object of his mathesis precisely this concept of quantity in general, 'magnitudo in genere', which applies to everything extended, whether countable or measurable. Thus does the abstract concept of 'universal quantity' – seen with the 'eye of the mind'– receive a clear foundation in Cartesian thinking: on the one hand, in operations of the translatability of geometry and arithmetic, and on the other hand in the

---

[27] On the publishing history see Descartes (1902, 351-396).
[28] Descartes (1902, 359-369).
[29] Rule 4, Descartes (1904, 376).
[30] This, incidentally, becomes the core concept of the idea of method in Descartes' epistemology.

representability of quantifiable objects in the extensional figurative language of the mathesis universalis. The 'magnitudo in genere' is created in the act of being brought to intuition.[31]

(4) A last step remains to be taken, since it can be shown that the attributes of the material world, the 'res extensa' (corporeal substance), are originally derived from the model of figurative extension. In the mathesis universalis of the early Descartes, 'figura et extensio' remain attributes of the symbolic system of depicting the material world; but in the classic 'metaphysical Descartes', figure and extension are promoted to fundamental traits of material reality itself. In the 'Regulae', the early Descartes initially grounds the privileged significance of 'figura' and 'extensio' in an argument rooted in his theory of perception.[32] 'Figure' is that which can be touched and seen, that which can be easily conveyed and that which is so general and simple that it "is found in every object of sensory perception"[33](Rule 12). All qualitative differences in perception (odours, colours, sounds) imprint themselves upon our optical apparatus in the form of two-dimensional graphic configurations. There is a sort of drawing pencil in the apparatus of perception. Descartes illustrates this reduction of qualities of perception to two-dimensional graphic figures by using three differently drawn rectangles, whose figurative differences are analogous to the differences in colour between white, blue and red.

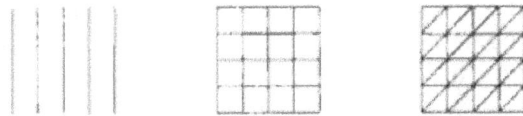

Fig. 5 The Figures Drawn in Rule 12 to Designate Differences in Colour[34]

Let us now summarize our diagrammatologic reconstruction: (i) The diagram of a coordinate system does not illustrate anything, but is rather a constructive tool of demonstrating and problem solving. It opens up a visual as well as tactile space to operate and manipulate abstract, non-spatial entities made visible. What matters here is the operative functioning of the graphism of coordinates. The meaning of a diagram is in its technical use. (ii) The interplay of line and surface constitutes a space with a twofold

---

[31] More precisely, see Krämer (1991, 201ff.).
[32] Rule 12, Descartes (1904, 412ff.)
[33] Ibid. 413.
[34] Descartes, Regulae ad directionem ingenii-Rules for the direction of the natural intelligence, a bilingual edition, ed. G. Heffernan, Amsterdam-Atlanta: Rodopi, 1998, 142.

character: a space for 'observing' and handling visual objects and a space for thinking about abstract objects. By virtue of this twofold nature, non-visual objects and theoretical entities can be spatialized, made intuitable and manipulable. This intermediary space corresponds to Plato's third subsection of the line. (iii) The coordinate system fulfils the function of mediating and translating between heterogeneous spheres in two senses: first, it mediates between figure and formula, geometry and arithmetic, and thus gives rise to the concept of universal quantity; second, it connects intuiting with thinking without relinquishing the differentiation between them.

The line as a field of research and discourse

So we see that amidst philosophical reflections in antiquity and the Early Modern Era, we come upon a form of reasoning that makes use of the 'cognitive power of the line'. Numerous scholars have already engaged with the culture-endowing role of the line.

In the field of cultural history, for Tim Ingold[35] the production and pursuit of lines is ubiquitous in human activities such as walking, weaving, storytelling, singing, drawing and writing. Ingold develops a comparative anthropology of the line, in which he tracks the transformations of the line in concurrence with traces, threads and surfaces. Threads out of which surfaces can be formed and traces that are imprinted on solid surfaces constitute the two basic modalities of his archaeology of the line. The interaction of thread and trace, culminating in notational systems, shows itself to be the cohesion of weaving and writing, of texture and text.

In the field of palaeontology, André Leroi-Gourhan[36] examines the function of graphism: the early linear forms of ornament, hunting symbol and record-keeping arise from a graphic aptitude which Leroi-Gourhan sets alongside speech as equally fundamental in its world-constituting function. Language and graphism are not only on an equal par in their cognitive potential. Rather, the drawing and reading of symbols, in contrast to acoustic signal formation, was not practised anywhere before the emergence of homo sapiens.[37] Leroi-Gourhan detaches the graphic from the artistic image and the pictures of the arts. In its evolutionary origin graphism is associated with abstraction, not with mimetic concretion.

In the field of art history Wolfgang Kemp reconstructs the history of the concept 'disegno' in the sixteenth century. 'Disegno' proves to be a

---

[35] Ingold (2007).
[36] Leroi-Gourhan (1980).
[37] Leroi-Gourhan (1989, 238).

double-sided concept: it means both the graphé as everyday "origin and outset of all human activity" and the non-everyday genius of a divine plan inherent in nature. 'Disegno', therefore, "mediates actively between nature and art"[38].

The art historian Horst Bredekamp draws upon Galileo's drawings of the moon, Darwin's coral-like diagrams of evolution, Mach's eye diagram, and Crick's spirals to analyse the cognitive power of the line: "On the border between thinking and materializing, these drawings and diagrammatic lines show a suggestive force all their own, which no other form of expression possesses... as the first trace of the corporeal on paper they embody thinking in its highest possible immediacy."[39] Bredekamp finds the line so essential to modern art and science precisely because it achieves a balance between intuiting and thinking, and thus – as previously set out in the term 'disegno'– plays a role in the sphere of the sensory as well as in that of the intellectual.

In the study of literature Georg Witte[40] offers a phenomenology of linear form and graphism in the interplay between the fluid performance of the writing or drawing hand and the contrasting stable figuration of lines. Witte examines the line in its double function: it is a heteropoetic medium of depiction and representation as well as an autopoetic creative power.[41] Between the line as an evidence-creating form of knowledge and as an aesthetic absolute lies a broad area, which artistic avant-gardes explore in practice and sound out in theory.

In cognitive semantics George Lakoff[42] and Mark Johnson[43], among others, demonstrate the ways in which spatial schemata metaphorically transfer bodily movements into cognitive domains. Thus does an implicit 'cognitive topology'[44] emerge, which reveals a nucleus of spatial orientation precisely in non-spatial, abstract thinking and in interactions with quantities. Alongside the 'container', 'interior/exterior', and 'peripheral/central', the line proves to be a central topological scheme of classification – in conjunction with the 'path' as the way between a point of departure and a destination. Can we then view the drawn diagram as the exteriorization of our mental inner topology, which in turn results from an archaic metaphorization of originally external corporeal activities in space?

---

[38] Kemp (1974, 227); in contemporary theory of art there have, however, been attempts at a 'non-fixing' concept of the line: Busch (2007); Derrida (1997); Elkins (1995); Rosand (2002).
[39] Bredekamp (2002, 24).
[40] Witte (2007); on the phenomenological approach to the line see also Lüdeking (2006).
[41] Witte (2007, 37).
[42] Lakoff (1988 and 1990).
[43] Johnson (1987).
[44] Ibid. 29.

These cursory notes should suffice for us to pose a question: is it possible to construct an 'epistemology of the line' in the sense of a rehabilitation and reconstruction of the explicit and implicit diagrammatic structures of rational thinking? This question outlines the research programme of 'Diagrammatology'. In continuation and revision of the present debate, this research programme seeks to shift two emphases:

(i) Frederick Stjernfelt[45] rehabilitated diagrammatical reasoning in his course-setting monograph 'Diagrammatology', tracing the field back to the thinking of Charles Sanders Peirce with further reference to Edmund Husserl. Though Peirce explicitly shaped the concept of 'diagrammatical reasoning', he is by no means its starting point: Sciences and philosophy have in fact been marked throughout history since their inception by diagrammatic dimensions, which remain to be reconstructed in the work of Leibniz, Lambert, Kant and, alongside Pierce, of Wittgenstein and Deleuze. In particular, Derrida's contribution must be acknowledged if 'Grammatology' is to be broadened into 'Diagrammatology'.

(ii) In 'The Philosophical Status of Diagrams', Marcus Greaves[46] sees the diagrammatical as historically fighting a rearguard action; it is, so to say, the loser on the battlefield of mathematical and logical strategies of symbolization. This interpretation is bound to Greaves' placement of figure and formula, and ultimately also image and language, in binary opposition to each other. But as soon as formalism is grasped in its notational iconicity and as a hybrid between language and image, it forms a complementary diagrammatic modality to the intuitive figure. The diagrammatic then undergoes modification; it is not, however, displaced.

The remaining space shall only allow us to point out a few cognitive-technical und epistemological aspects of 'diagrammatics in the spirit of an epistemology of the line'.

'Epistemology of the line': Facets of a theory of diagrammatics

(1) Simultaneity: We assume a fundamental exteriority of the human mind,[47] which holds a diagrammatic dimension.[48] This dimension includes

---

[45] Stjernfelt (2007).
[46] Greaves (2002).
[47] Koch and Krämer (2009).
[48] Also arising here is a connection to the idea of the 'extended mind', Clark and Chalmers (1998), and of the 'embodies and embedded mind', Clark (2008).

the use of notations, schemes, charts, tables, diagrams and maps for both practical and theoretical purposes. Regardless of the variety among these forms of depiction, they all share an essential attribute: they employ two-dimensional spatial configurations as the matrix and medium to depict theoretical matters and 'objects of knowledge'. In contrast to the sequentiality of auditory and tactile impressions, seeing is grounded in simultaneity.[49] When our eyes are presented with things that are next to each other, we gain an overview; we can compare different sorts of things to see similarities and differences, as well as recognize relationships, proportions and patterns within the vast diversity. Diagrams make the dissimilar comparable.[50] Let us consider the Simile of the Divided Line: it is a single line through which Plato visualizes diametrically opposed realms of being. And it is the trick of the Cartesian use of the coordinate system to make the measurable and the countable mutually transferable. The homogenization of the heterogeneous—incidentally, a root of all concept formation – is a key to the potential of the diagrammatic 'handwork of the mind.'

(2) Own spatiality: This homogenizing function, which mediates between divergent things,[51] is possible to the extent that diagrammatic inscriptions span their 'own spatiality'. This space exhibits three essential traits: it is (i) a two-dimensional surface, for the most part conceived without simulation of depth (e.g. central perspective); (ii) it is formatted and scaled, and thus evinces a directedness and a system of measure; (iii) it is composed as an interplay of inscription and blank space. All of this arises from the interaction of point, line and plane, the simplest product of which the schema[52] is: schemata here understood as the possibility of presenting the shapeless as a shape. Lines constitute the archetypal form of clear shape-forming: they delimit and they exclude. Every mark on a surface creates an asymmetry that becomes the source of potential distinction: the circular line separates points within and outside of the circle; a line segment separates what is left or right of it, above or below it.

---

[49] Jonas (1997, 248ff.).
[50] For Jonas (1997, 259), seeing is thus among the five senses the privileged one for objectivity.
[51] Cf. Krämer (2008) on the role of a 'third party' as an entity of transmission and mediation, residing between heterogeneous systems as a fundamental definition of 'mediality'. There is thus a genuine connection between media theory and theory of the diagrammatic.
[52] It is thus no accident that Immanuel Kant's 'transcendental schematism' (B 176 - B187) is illustrated through the example of the drawing of a line, and that he accentuates the line's double character as "on the one hand being intellectual, on the other hand sensory" (B 178).

Unlike the line, the point as an elementary form of the graphic is little investigated.[53] The question of whether and how points are inserted and interpreted separates eras of civilization: we need only consider the laying out of numbers with the aid of calculi in Pythagorean pebble (psephoi) arithmetic, or the zero as the originating point of the coordinate system, or the end point reached in an oscillating movement before the direction of motion is reversed, or the vanishing point in the construction of perspective, or the dot shape of musical notes, or the punctuation marks used in many notational systems.

(3) Hybridity: Diagrammatic inscriptions join together traits of the iconic and the linguistic. They are characterized by the fusion of showing and telling.[54] In contrast to artistic images and spoken language they take a middle position in-between the two. The presumption of a disjunctivity between 'pure' image and 'pure' language proves phenomenologically mistaken: 'image' and 'language', iconicity and discursivity, are to be differentiated only as concepts; they form the opposite ends of a conceptual scale, between which all real symbolic phenomena reside in their respective mixtures of the two. In relation to diagrammatics, 'iconicity' has a threefold meaning: two-dimensionality, visuality, and trans-naturalistic resemblance. Discursiveness emphasizes three further attributes: syntacticity, referentiality and propositionality. To put it bluntly: diagrams are images that make assertions, and thus can be right or wrong.

In modern theories of the image, it was standard practice[55] to emphatically reject resemblance as a definiens of iconicity. But to the extent that diagrammatic inscriptions show what they tell, the principle of structural resemblance is fundamental to them; 'structural resemblance' in the same sense in which the formula of a circle resembles its concrete figure or the number of contour lines on a topographical map corresponds to the measured height of a mountain.[56]

(4) Referentiality: Unlike the artistic image, which in the first line refers to itself, diagrammatic artefacts require an external point of reference: something is being shown. Diagrammatic inscriptions are transcriptions[57] and in this sense manuals or machines of translation. Referentiality,

---

[53] Schäffner (2003) is an exception.
[54] In all languages, however, showing is fundamental, just as images, conversely, are imbedded in texts.
[55] e.g. Goodman (1997).
[56] On this concept of a 'trans-natural illustration', see Krämer (2008, 311).
[57] On the theory of transcriptivity as constituting the transcribed, see Jäger (2002).

however, must not be naively understood as reference to a matter that exists independent of inscription. There are two reasons for this: (i) Only through the inner logic of diagrammatic spatiality is it possible to depict anything in a standardizing way, and that which is depicted is in this respect constitutively shaped by the medium. (ii) The 'objects' of diagrammatic depiction are always relations and proportions, which are not 'inherent' but are created by intellectual practices in the interaction of eye, hand and mind. Even topographical maps do not simply depict a landscape, but rather a knowledge of a landscape, according to the methods of projection that necessarily distort the three-dimensional.

(5) Operativity: Notations, diagrams and maps do not merely depict something; rather, they are used every day and for the most part unspectacularly, as cultural techniques. This aspect of their use shows a cognitive and a communicative dimension: (i) In cognitive terms, diagrammatic depictions open up a two-dimensional space for handling, observing and exploding the depicted. Whether sailors use a map to chart their course or chemists discover gaps in the inscriptional framework of chemical elements,[58] whether a musical score enables pieces of music to be analysed or composed in new ways, whether we solve arithmetic equations with the aid of algebraic formulas[59] or create artistic and scholarly texts through the formulation, deletion and rewriting of sentences:[60] the inscribed surface always creates a space of operations and the diagrammatic always functions as a tool and instrument of orientation, analysis, revision and reflection. (ii) In communicative terms, diagrammatic artefacts are, in their generally manageable format, thoroughly mobile. Not only do they serve to make something perceptible; that which is depicted can also be effortlessly transmitted, transported, circulated, reproduced and combined. As Bruno Latour puts it: "Inscriptions are not interesting per se but only because they increase the mobility or the immutability of traces."[61]

(6) Dependence: No line interprets itself.[62] It is precisely Plato's Simile of the Divided Line and the history of its reception that shows us so starkly that the same matter can be visualized diagrammatically in a variety of ways and, conversely, that the same diagram can elicit thoroughly divergent interpretations. Moreover, the meaning of historically transmitted

---

[58] Klein (2003 and 2005).
[59] Krämer (1988).
[60] Raible (2004).
[61] Latour (1990, 31).
[62] Lüthy and Smets (2009).

diagrammatic artefacts is scarcely to be understood apart from the socio-cultural milieu of their usage or without situating them in a 'tacit knowledge' of graphic conventions and their interpretation. The diagrammatic is situated: whether in the context of the text in which it appears or – here consider architectural drawings and city plans – within practices. It comes as no surprise that the diagrammatic is in most cases a part of a cascade of transmissions and transcriptions.[63]

References

Anderson, Michael, Meyer, Berndt, and Olivier, Patrick (eds.), Diagrammatic Representation and Reasoning (Berlin: Springer, 2001).

Aristotle, Metaphysics (Sioux Falls: Nu Vision Publications, 2005).

Boehm, Gottfried (ed.), Was ist ein Bild? (Munich: Fink, 1994).

―― 'Die Wiederkehr der Bilder', in id. (ed.), Was ist ein Bild? (Munich: Fink, 1994), 11-38.

Bogen, Steffen, 'Schattenriss und Sonnenuhr. Überlegungen zu einer kunsthistorischen Diagrammatik', Zeitschrift für Kunstgeschichte, 68/2 (2005), 153-176.

―― and Thürlemann, Felix, 'Jenseits der Opposition von Text und Bild. Überlegungen zu einer Theorie des Diagramms und des Diagrammatischen', in Alexander Patschovsky (ed.), Die Bildwelt der Diagramme Joachims von Fiore. Zur Medialität religiös-politischer Programme im Mittelalter (Ostfildern: Thorbecke, 2003), 1-22.

Bredekamp, Horst, 'Die Erkenntniskraft der Linie bei Galilei, Hobbes und Hooke', in Barbara Hüttel (ed.), Re- Visionen. Zur Aktualität von Kunstgeschichte (Berlin: Akademie, 2002), 145-160.

―― and Schneider, Pablo (eds.), Visuelle Argumentationen. Die Mysterien der Repräsentation und die Berechenbarkeit der Welt (Paderborn: Fink, 2006).

―― Fischel, Angela, Schneider, Birgit, and Werner, Gabriele, 'Bildwelten des Wissens', in Bilder in Prozessen. Bildwelten des Wissens. Kunsthistorisches Jahrbuch für Bildkritik, 1 (Berlin: Akademie, 2003), 9-20.

---

[63] Latour (1990).

Brentlinger, John A., 'The Divided Line and Plato's "Theory of Intermediates"', Phronesis, 8 (1963), 146-166.

Busch, Werner, 'Die Möglichkeiten der nicht-fixierenden Linie. Ein exemplarischer historischer Abriß', in Wilhelm Busch, Oliver Jehle, and Carolin Meister (eds.), Randgänge der Zeichnung (Paderborn: Fink, 2007), 121-140.

—— Jehle, Oliver and Meister, Carolin (eds.), Randgänge der Zeichnung (Paderborn: Fink, 2007).

Châtelet, Gilles, Figuring Space: Philosophy, Mathematics and Physics (Dordrecht: Kluwer, 2000).

Clark, Andy, 'Pressing the Flesh: Exploring a Tension in the Study of the Embodied, Embedded Mind', Philosophy and Phenomenological Research, 76/1 (2008), 37-59.

—— and Chalmers, David, 'The Extended Mind', Analysis, 58/1 (1998), 7-19.

Daston, Lorrain, and Galison, Peter, 'Das Bild der Objektivität', in Peter Geimer (ed.), Ordnungen der Sichtbarkeit. Fotografie in Wissenschaft, Kunst und Technologie (Frankfurt am Main: Suhrkamp, 2002), 29–99.

—— Objektivität (Frankfurt am Main: Suhrkamp, 2007).

Derrida, Jacques, Aufzeichnungen eines Blinden. Das Selbstportrait und andere Ruinen, ed. Michael Wetzel (Munich: Fink, 1997).

Descartes, René, Oeuvres de Descartes, ed. Charles Adam and Paul Tannery, vol. vi (Paris: Leopold Cerf, 1902), 367-485.

—— 'Regulae ad directionem ingenii', in Charles Adam and Paul Tannery (eds.), Oeuvres de Descartes, vol. x (Paris: Leopold Cerf, 1904).

—— The Geometry of Rene Descartes, trans. David Eugene Smith and Marcia Latham (New York: Cosimo, 2007).

Earnshaw, Rae, and Wiseman, Norman, An Introductory Guide to Scientific Visualization (Berlin: Springer, 1992).

Elkins, James, 'Marks, Traces, "Traits", Contours, "Orli", and "Splendores": Nonsemiotic Elements in Pictures', Critical Inquiry, 21/4 (1995), 822-860.

Fogelin, Robert J., 'Three Platonic Analogies', Philosophical Review, 80 (1971), 371-382.

Fritz, Kurt von, 'The Discovery of Incommensurability by Hippasus of Metapontum', The Annals of Mathematics, 46/2 (1945), 242-264.

Gehring, Petra et al., Diagrammatik und Philosophie (Amsterdam: Rodopi, 1992).

Goodman, Nelson, Sprachen der Kunst. Entwurf einer Symboltheorie (Frankfurt am Main: Suhrkamp, 1997).

Gormans, Andreas, 'Imaginationen des Unsichtbaren. Zur Gattungstheorie des wissenschaftlichen Diagramms', in Hans Holländer (ed.), Erkenntnis, Erfindung, Konstruktion. Studien zur Bildgeschichte von Naturwissenschaften und Technik vom 16. bis zum 19. Jahrhundert (Berlin: Mann, 2000), 51-71.

Greaves, Mark, The Philosophical Status of Diagrams (Stanford: CSLI Publications, 2002).

Grube, Gernot, Kogge, Werner, and Krämer, Sybille (eds.), Schrift. Kulturtechnik zwischen Auge, Hand und Maschine, Reihe Kulturtechnik (Munich: Fink, 2005).

Halawa, Mark Ashraf, Wie sind Bilder möglich? (Cologne: Halem, 2008).

Heintz, Bettina, and Huber, Jörg (eds.), Mit dem Auge denken. Strategien der Sichtbarmachung in wissenschaftlichen und virtuellen Welten (Zurich: Edition Voldemeer, 2001).

Heßler, Martina, 'Annäherungen an Wissenschaftsbilder', in id. (ed.), Konstruierte Sichtbarkeiten. Wissenschafts- und Technikbilder seit der frühen Neuzeit (Munich: Fink, 2006), 11-37.

Ingold, Tim, Lines: A Brief History (London/New York: Routledge, Chapman & Hall, 2007).

Jäger, Ludwig, 'Transkriptivität. Zur medialen Logik der kulturellen Semantik' (2002), in Ludwig Jäger and Georg Stanitzek (eds.), Transkribieren (Munich: Fink, 2007), 19-41.

Jamnik, Mateja, Mathematical Reasoning with Diagrams (Stanford: CSLI Publications, 2001).

Johnson, Mark, The Body in the Mind: The Bodily Basis of Meaning, Imagination and Reason (Chicago: University of Chicago Press, 1987).

Jonas, Hans, 'Der Adel des Sehens', in Ralf Konersmann (ed.), Kritik des Sehens (Leipzig: Reclam, 1997), 247-271.

Kemp, Wolfgang, 'Disegno. Beiträge zur Geschichte des Begriffs zwischen 1547 und 1607', Marburger Jahrbuch für Kunstwissenschaften, 19 (1974), 218-240.

Klein, Ursula, Nineteenth-Century Chemistry – Its Experiments, Paper Tools and Epistemological Characteristics (Stanford: Stanford University Press, 2003).

—— 'Visualität, Ikonizität, Manipulierbarkeit: Chemische Formeln als "paper tools"', in Gernot Grube, Werner Kogge and Sybille Krämer (eds.), Bild, Schrift, Zahl (Munich: Fink, 2005), 237-252.

Koch, Peter, and Krämer, Sybille (eds.), Schrift, Medien, Kognition. Über die Exteriorität des Geistes, 2nd ed. (Tübingen: Stauffenburg, 2009).

König, G., 'Theorie', in Joachim Ritter (ed.), Historisches Wörterbuch der Philosophie, vol. 10 (Darmstadt: Wissenschaftliche Buchgesellschaft Darmstadt, 1998), 1128-1146.

Krämer, Sybille, Symbolische Maschinen (Darmstadt: Wissenschaftliche Buchgesellschaft Darmstadt, 1988).

—— 'Über das Verhältnis von Algebra und Geometrie in Descartes' Géomètrie', Philosophia Naturalis, 26/1 (1989), 19-40.

—— Berechenbare Vernunft. Kalkül und Rationalismus im 17. Jahrhundert (Berlin/New York: de Gruyter, 1991).

—— and Bredekamp, Horst, Bild Schrift Zahl, Reihe Kulturtechnik (Munich: Fink, 2003, 2nd ed. 2008).

—— 'Writing, Notational Iconicity, Calculus: On Writing as a Cultural Technique', Modern Language Notes – German Issue, John Hopkins University Press, 118/3 (2003), 518-537.

—— 'Operationsraum Schrift. Ein Perspektivwechsel im Schriftverständnis', in Gernot Grube, Werner Kogge and Sybille Krämer (eds.), Schrift. Kulturtechnik zwischen Auge, Hand und Maschine, Reihe Kulturtechnik (Munich: Fink, 2005), 13-32.

—— Medium, Bote, Übertragung. Kleine Metaphysik der Medialität (Frankfurt am Main: Suhrkamp, 2008).

—— 'Operative Bildlichkeit. Von der 'Grammatologie' zu einer 'Diagrammatologie'? Reflexionen über erkennendes Sehen', in Martina Hessler and Dieter Mersch (eds.), Logik des Bildlichen. Zur Kritik der ikonischen Vernunft (Bielefeld: Transcript, 2009), 94-123.

—— and Koch, Peter, (eds.), Schrift, Medien, Kognition. Über die Exteriorität des Geistes, 2nd ed. (Tübingen: Stauffenburg, 2009).

Lakoff, George, 'Cognitive Semantics', in Umberto Eco, Marco Santambrogio and Patrizia Violi (eds.), Meaning and Mental Representations (Bloomington: Indiana University Press, 1988), 119-154.

—— 'The Invariant Hypothesis: Is abstract reason based on image-schemas?', Cognitive Linguistics, 1/1 (1990), 39-74.

Latour, Bruno, 'Drawing Things Together', in Michael Lynch and Steve Woolgar (eds.), Representation in Scientific Practice (Cambridge: MIT Press, 1990), 19-68.

Leroi-Gourhan, André, Hand und Wort. Die Evolution von Technik, Sprache und Kunst (Frankfurt am Main: Suhrkamp, 1980).

Lüdeking, Karlheinz, 'Bildlinie/Schriftlinie', in id., Die Grenzen des Sichtbaren (Munich: Wilhelm Fink, 2006).

Lüthy, Christop, and Smets, Alexis, 'Words, Lines, Diagrams, Images: Towards a History of Scientific Imagery', Early Science and Medicine, 14 (2009), 398-439.

Lynch, Michael, 'The Production of Scientific Images: Vision and revision in the history, philosophy and sociology of science', Communciation & Cognition, 31/2/3 (1998), 213-228.

May, Michael, 'Diagrammtisches Denken. Zur Deutung logischer Diagramme als Vorstellungsschemata bei Lakoff und Peirce', Zeitschrift für Semiotik, 17/3-4 (1995), 285-302.

Mersch, Dieter, 'Visuelle Argumente. Zur Rolle der Bilder in den Naturwissenschaften', in Sabine Maasen, Thorsten Mayerhauser and Cornelia Renggli (eds.), Bilder als Diskurse – Bilddiskurse (Weilerswist: Velbrück Wissenschaft, 2006), 95-116.

Mitchell, William J., 'The Pictorial Turn', Art Forum, 30/7 (1992), 89-94.

—— Picture Theory: Essays on Verbal and Visual Representation (Chicago: University of Chicago Press, 1994).

Natopoulos, James Anastasios, 'Movement in the Divided Line of Plato's Republic', Harvard Studies in Classical Philology, 47 (1936), 57-83.

Plato, Republic, trans. Robin Waterfield (Oxford/New York: Oxford University Press, 2003), 238-239.

Raible, Wolfgang, 'Über das Entstehen der Gedanken beim Schreiben', in Sybille Krämer (ed.), Performativität und Medialität (Munich: Fink, 2004), 191-214.

Rosand, David, Drawing Acts: Studies in Graphic Expression and Representation (Cambridge: Cambridge University Press, 2002).

Rottmann, Michael, 'Das digitale Bild als Visualisierungsstrategie der Mathematik', in Ingeborg Reichle, Steffen Siegel and Achim Spelten (eds.), Verwandte Bilder. Die Fragen der Bildwissenschaft (Berlin: Kulturverlag Kadmos, 2008), 281-296.

Schäffner, Wolfgang, 'Punkt. Minimalster Schauplatz des Wissens im 17. Jahrhundert (1585-1665)', in Helmar Schramm, Ludger Schwarte, and Jan Lazardig (eds.), Kunstkammer, Laboratorium, Bühne. Schauplätze des Wissens im 17. Jahrhundert, Theatrum Scientiarum, vol. 1 (Berlin/New York: de Gruyter, 2003), 56-74.

Schmidt-Burkhardt, Astrit, Stammbäume der Kunst. Zur Genealogie der Avantgarde (Berlin: Akademie, 2005).

—— 'Wissen als Bild. Zur diagrammatischen Kunstgeschichte', in Martina Hessler and Dieter Mersch (eds.), Logik des Bildlichen. Zur Kritik der ikonischen Vernunft (Bielefeld: Transcript, 2009), 163-187.

Siegel, Steffen, Tabula. Figuren der Ordnung um 1600 (Berlin: Akademie, 2009).

Smith, Nicholas D., 'Plato's Divided Line', Ancient Philosophy, 16/1 (1996), 25-47.

Stekeler-Weithofer, Pirmin, Formen der Anschauung (Berlin/New York: de Gruyter, 2008).

Stjernfelt, Frederik, 'Diagrams as Centerpiece of a Peircian Epistemology,' Transactions of the Charles S. Peirce Society, 36/3 (2000), 357-384.

—— Diagrammatology: An Investigation on the Borderline of Phenomenology, Ontology and Semiotics (Berlin: Springer, 2007).

Stocks, J.L., 'The Divided Line', Classical Quarterly, 5 (1911), 72-88.

Tufte, Edward, Visual Explanations – Images and Quantities, Evidence and Narrative (Cheshire: Graphics Press, 1997).

Ueding, Wolfgang Maria, 'Die Verhältnismäßigkeit der Mittel bzw. Die Mittelmäßigkeit der Verhältnisse: Das Diagramm als Thema und Methode der Philosophie am Beispiel Platons, bzw. Einiger Beispiele Platons', in Petra Gehring et al., Diagrammatik und Philosophie (Amsterdam/Atlanta: Rodopi, 2002).

Witte, Georg, 'Die Phänomenalität der Linie – graphisch und graphematisch', in Wilhelm Busch, Oliver Jehle, and Carolin Meister (eds.), Randgänge der Zeichnung (Paderborn: Fink, 2007), 29-54.

Yang, Moon Heum, 'The relationship between hypothesis and images in the mathematical subsection of the divided line in Plato's Republic', Dialogue – Canadian Philosophical Review, 44/2 (2005), 285-312.

Zittel, Claus, 'Abbilden und Überzeugen bei Descartes', in Karl A.E. Enenkel and Wolfgang Neuber (eds.), Cognition and the Book – Typology of Formal Organisation of Knowledge in the Printed Book of the Early Modern Period (Leiden: Brill, 2005), 535-601.

Space, Structure, and Similarity.
On Representationalist Theories of Diagrams

by
Jan Wöpking

Abstract

Representationalist appraoches view diagrams as representations with specific properties. I examine the two-fold claim that diagrams are (i) two-dimensional graphical representing vehicles that (ii) are structurally similar to what they represent. Drawing on work by Palmer and Shimojima I argue that this claim misses an important kind of similarity, namely nomic similarity. I argue that it is precisely nomic similarity that accounts to a large extent for the cognitive power of diagrams. After introducing the concept of nomic similarity I propose a taxonomy of diagram definitions that is organized around two key distinctions. The first distinguishes between broad and narrow definitions of diagrams. Narrow definitions restrict the notion of the diagram to two-dimensional graphical representing vehicles, broad definitions do not have such a restriction. The second distinction is between definitions that pay attention to the importance of nomic similarity and those that do not.

1. Introduction

In this paper I discuss different ways of defining diagrams. I will focus on representationalist accounts of diagrams. In a representationalist perspective diagrams are taken to be representations with specific properties. The approach I would like to discuss here consists in two logically independent claims that taken together constitute a promising attempt at defining diagrams. The first claim concerns the relationship between a diagram and what it represents. It proposes that diagrams are representing vehicles that stand in a relation of similarity to the object they represent. This similarity, however, is not based upon perceptual or physical but on structural similarity. A diagram may not look like the entitity it represents but can still be similar in that it has the same abstract or logical structure as its object. This condition does not in itself suffice to single out the entities we would like to identify as diagrams, because a lot of representations may be structurally similar to their respective objects. Thus, a second claim is required. It concerns the physical medium of the diagram. In this view diagrams are specified as spatially structured two-dimensional (schematic) graphics. The property of being spatially structured means that diagrams are structured in way that allows for geometrical or topological interpretation. If combined, the two claims propose the following criterion for a diagram: A diagram is (i) a representing vehicle that stands in a relation of structural similarity to its object and (ii) this structural similarity is implemented or realized by a spatial structure in a two-dimensional graphical medium. In this definition of a diagram a special form of mapping and a special form of medium come together. I would like to call it the Triple-S-Definition of diagrams because it takes a diagram to be a spatially realized representational vehicle that is structurally similar to its object.[1]

This paper aims at discussing the Triple-S-Definition. Drawing on work by Palmer and Shimojima I argue that this claim misses an important kind of similarity, namely nomic similarity. I argue that it is precisely nomic similarity that accounts to a large extent for the cognitive power of diagrams. After introducing the concept of nomic similarity I propose a taxonomy of diagram definitions that is organized around two key distinctions. The first distinguishes between broad and narrow definitions of diagrams. Narrow definitions restrict the notion of the diagram to two-dimensional graphical representing vehicles, broad definitions do not have such a restriction. The second distinction is between definitions that pay attention to the importance of nomic similarity and those that do not.

2. Representations

In a representationalist perspective diagrams are said to be representations. But what is a representation? There are literally hundreds of different answers to this question. Worse, the notion of representation itself has come under severe attack in the last century. It was considered useless, even harmful. This is especially true for attempts to develop a kind of picture theory of representation, i. e. a theory that grounds representation in similarity. I will not enter these debates here. For my purposes a simplified, mathematically inspired notion of representation shall suffice. In this simple form a representation is considered a triple consisting in (i) a represented

---

[1] This proposal is quite popular. It exists in various versions, each employing its own terminology, its own style of formalization and so on. But these are superficial differences. The Triple-S-Definition is widely discussed in recent German debates on a science of diagrams which is often called Diagrammatik, sometimes also Diagrammatologie. See Krämer (2009) for an attempt at terminological clarification.

world ("object", "target"), (ii) a representing world ("representing vehicle", "sign", "source"), and (iii) a mapping between the worlds.[2]

Instead of "worlds" one could also speak of situations, domains, or structures. A world is just a complex, structured entity, consisting in elements, properties of the elements and relations on the elements. Representation means that one world is somehow mapped on another world. Mapping means that elements, properties and relations of one world are partially or completely paired with elements of the other world. It correlates one world with another. Usually only parts of a world are represented and usually only parts of a world represent. In both cases there have to be rules to specifiy which parts of a world are relevant for the representation and which are not.

The representational theory claims that diagrams are representations with specific properties. This can involve each of the three parts of the representating situation. In this paper I will concentrate on (ii) and (iii) only.

3. Physical similarity

This is probably the most undisputed claim when it comes to diagrams: They do not have to look like what they represent. Similarity in appearance is neither a sufficient nor a necessary condition for a diagram. The same goes for physical similarity. Examples for this claim are easily provided. A diagram of the heart mechanism neither looks like a heart nor is physically similar to one. A Galilean diagram representing a falling body does not look like a falling body. While often repeated, it has not always been made clear if the similarity in question was one of experienced similarity or one of objective similarity. Is the similarity objective in the sense that the diagram does in fact share visually detectable physical properties with the object it represents? Or does the similarity merely have to be an "experienced similarity" for the user? A diagram could, of course, be experienced as similar to something else without it actually being similar.

A good working definition of physical similarity is provided by O´Brien and Opie (2004). According to them a "representing vehicle and its object resemble each other at first order if they share physical properties, that is, if they are equal in some respects" (p. 9). Of course we have to state in which respects the representing vehicle and its object shall be similar. There can be many ways in which a physical similarity can exist depending on

---

[2] There is an obvious ambiguity in the way the term representation is used. It can either refer to the triple or only to the representing vehicle. Usually context determines easily which one is meant.

what you unterstand by "physical property". Something that is green can be similar to some other thing in being green, in being of such and such size, in being of such and such form etc. When stated like this it becomes obvious that this similarity cannot be a defining property of many things we would like to identify as diagrams. It is, however, important to note that the no-physical-similarity claim does not entail that a diagram must be physically dissimilar to its object. We can think of cases in which a diagram was much easier to understand if it preserved the colors of things it represented. It seems (at least at first sight) to be unreasonable to depict the presence of a lake in a map by a green color and the presence of a forest by a blue color.[3] In these cases, though, the no-physical-similarity claim can be upheld by proposing that the diagram could be transformed into a different kind of diagram that was informationally equivalent but exhibited no physical similarity to its object. In the case of the map an inversion of the colors would be possible. This is to say that physical similarity is not a necessary condition for a diagram but a contingent fact. It might make it easier to access information but it is not a logical feature of diagrams.

4. Stuctural similarity

If it is not a similarity in terms of physical properties or in terms of appearances, what kind of similarity does characterize a diagram? In german papers a diagram is sometimes called a Strukturbild (structural image) that bears a Strukturähnlichkeit to its object. This points to the answer. Diagrams are said to be structurally similar to their objects. Peirce is often and rightly considered a main proponent of this idea. Here is a very clear and exemplary passage: „Many diagrams resemble their objects not at all in looks; it is only in respect to the relations of their parts that their likeness consists" (CP 2.282).[4] As has also been repeatedly remarked, the most remarkable aspect about Peirce's account is his ingeneous way of defining similarity. Peirce writes: "For a great distinguishing property of the icon is that by the direct observation of it other truths concerning its object can be discovered than those which suffice to determine its construction" (CP 2.279). As Stjernfelt (2006) has pointed out the clue of this formulation is that "the decisive test

---

[3] Gombrich (1984) makes a similar remark: "We would be puzzled to find a map of London in which the parks were marked blue and the ponds green, because the other arrangement is so much easier to learn and keep in mind" (p. 184).

[4] Following convention, references to Peirce are are to volume and paragraph of: Peirce, C. S. 1960. Collected Papers. Ed. Charles Hartshorne and Paul Weiss. 8 vols. Cambridge, MA: Belknap P.

for iconicity lies in whether it is possible to manipulate or develop the sign so that new information as to its object appears" (p. 72).

In recent papers structural similarity is mostly defined in quasi-mathematcial terms as a kind of homomorphism or isomomorphism. O'Brien and Opie (2004) talk of second-order instead of first-order resemblance: "In second-order resemblance, the requirement that representing vehicles share physical properties with their represented objects can be relaxed in favour of one in which the relations among a system of representing vehicles mirror the relations among their objects" (p. 10). We can dinstinguish three forms such similarity can take: (i) a basic form: some elements and or some relations of a structure are preserved in the representing structure; (ii) a homomorphism: all elements and relations of a structure are preserved in the representing structure, although the converse is not necessarily possible. (iii) an isomorphism: all elements and relations of a structure are preserved in the representing structure, and the converse is also possible.

The idea of structural similarity can be further differentiated. Drawing upon Barwise and Hammer (1995) we can identify three criteria that a diagrammatic representation has to satisfy in order to be structurally similar to its object:

Correspondence of elements. Elements in the target domain have corresponding objects in the source domain. Diagrams usually obey a principle of icon identity, i. e. different tokens represent different objects. This feature constitutes a major difference to linguistic systems where it is common that different tokens can represent the same object.

Correspondence of relations. If the icon tokens stand in some relevant relationship RS, then there holds a relationship RT, represented by RS, among the objects that the icons denote. The converse holds as well.[5]

Correspondence of the structure of relations. (c) If a relationship RS among icon tokens has some structural property (such as transitivity, asymmetry, irreflexivity, etc.), then this same property must hold of the target relation RT represented by RS. The converse holds as well. This condition is of great importance and will be further discussed in the next two sections.

A common example of structural similarity is a geographical map. A map usually preserves metrical and or topological relations between the objects in the part of the world it depicts. If locations a, b, c stand in some relation in the world (say, a is closer to b than to c), then the corresponding tokens a', b', c' in the map stand in the same relation. The same goes for in-

---

[5] R_T refers to a relation in the target domain, i. e. the represented world. R_S referts to a relation in the source domain, i. e. the representing world or representing vehicle.

betweenness or connectedness or a variety of other kinds of relations. Euler diagrams mirror set inclusion by spatial topological inclusion. In chemistry, structural formulas represent atoms by letters, and the binary relation of two atom´s being connected by a smale line between the letters representing them.

5. Similarity of the Properties of Relations

The account of structural similarity, although interesting in itself, has to be further differentiated. The aim is to show that there is more than one way in which a structure-preserving mapping can take place, depending on the kind of representing relations and of the representing media. This idea has been explored by Palmer and by Shimojima.[6]

As quoted earlier, Palmer (1978) considers a representation to be a mapping between represented and representing world such that "at least some relations in the represented world are structurally perserved in the representing world" (pp. 266-267). This is apparently a version of structural similarity. However, Palmer does not stop at this point but goes on to distinguish two different modes of structure preservation. His presentation rests on the idea of an informational equivalence of two representations. This holds whenever two representations represent exactly the same set of objects and the same set of relations. Palmer then argues that there are at least two different ways in which representations can differ even if they preserve the same informational content. Basically the difference is this: A structure can be considered as consisting in elements plus relations on the elements. The relations in turn have certain properties, e. g. they can be transitive, symmetric, reflexive, anti-symmetric, asymmetric and so on. Now, the question is: Does the representing relation intrinsically have the same properties as the relation it represents? Or does the representing relation lack the properties in which case the structure has to be extrinsically enforced on the representation by a user. We define that intrinsic similarity is the case if the properties of the represented relation match the properties of the representing relation. Palmer writes: "Representation is (purely) intrinsic whenever a representing relation has the same inherent constraints as its represented relation. That ist, the logical structure required of the representing relation is intrinsic to the relation itself rather than imposed from outside" (p. 271). The intrinsic method is thus "to model a represented relation or dimension by using a representing relation or dimension that has the same inherent structure as that which it represents. In such cases, the

---

[6] Shimojima (2001) explicitly discusses Palmer´s work.

preservation of logical structure is a 'natural' consequence of the representing relation" (p. 296). If the representing relation does not match the represented relation via inherent logical properties, than the only way to preserve structure is by explicitly enforcing the structure on the elements. Let us consider a (rather artifical) example that Palmer gives to to get a better understanding of the idea.

The world to be represented consists in four objects each having a different size and the relation "is bigger than". This relation is asymmetric (if a is bigger than b, then b cannot be bigger than a) as well as transitive (if a is bigger than b, and b is bigger than c, than a is also bigger than c).[7] Consider an intrinsic representation, call it RI. RI represents this world by four lines of different length. On the other hand consider an extrinsic representation, call it RE, that consists in a directed graph made up of four nodes plus a number of edges. Two nodes are connected by a directed edge (an arrow) if the corresponding objects in the represented world stand in a "has bigger size than" relation. RI and RE represent the same structure but RI does this by means of intrinsic constraints, whereas RE uses extrinsic constraints. In RI the spatial structure of the graphical medium is constrained by geometrical laws in such a way as that the structure of the relations naturally holds. The longer-than relation exhibits the same logical properties as the bigger-than relation. It is easy to see that several other relations also qualify for such a matching, e. g. "shorter than", "larger than". It is, however, important to note that non-visual relations can also have the same logical properties, e. g. "brighter than" or "louder than" or "hotter than". They are all inherently asymmetric and transitive. In the case of RE however we have to explicitly represent the asymmetric and transitive nature of the relation. There are no intrinsic constraints doing this for us. We have to do it ourselves whereas in RI we only have to take care that the four elements are represented appropriately.

Intrinsic representations work in cases of multidimensional representations, too, as is made clear by the following example by Palmer. Let us assume we want to represent height, width and area of rectangles. Obviously there holds an intrinsic constraint between the three dimensions in such a way that the values of any two of the variables determine the value of the third (area = height * width). Let us again consider two different representations of the rectangles that are informationally equivalent. One of the representations shall be realized so that it was logically similar to the represented structure. Palmer suggests the following approach: "If the height

---

[7] It may have further properties but that does not need concern us here. The argument would essentially be the same.

of rectangles were modeled, say, by the volume of spheres, and if the width of rectangles were modeled by their density, then the area of rectangles would be intrinsically represented by their mass" (p. 274). Here again the values of two variable determine the value of the third variable (as mass = density * volume). That is why we do not have to represent the area explicitly. It is already implicitly represented once we have volume and densitiy of the spheres. The second representation represents height by length of line, width by brightness of the line and area by orientation of the line. In this case there is no natural constraint governing all three dimensions in a similar way as in the case of the height, width and area of the rectangles which are to be represented (though there is a matching between the individual constraints governing height and length and width and brightness). Therefore we have explicitly to provide for an accurate orientation of the line in order to get a true representation of the target world.

What is the use of Palmer´s notion for a theory of diagrams? Usually diagrams are defined by reference to structural instead of physical similarity only. Palmer´s paper points to a very important third kind of similarity. This similarity holds if a relation of a structure of an object is represented by a relation that has some or all of the logical properties of the represented relation. Palmer himself calls this "natural" similarity. It is based on the identity of the abstract logical properties of the relations holding on the elements. This kind of similarity lies between pure physical and pure structural similarity. Where exactly this kind of similarity is to be located is a question of degree, not of kind. On the one hand it is less similar than a physical similarity, one that preserves all or at least some relevant physical properties of the represented world. On the other hand it is more similar than a structural similarity that is preserved by means of extrinsic constraints only. Applied to diagrams it seems likely that it is precisely this third kind of similarity that is responsible for many of the typical properties that we encounter in diagrammatic reasoning. It will be discussed in greater detail in the next section.

6. Nomic similarity

Shimojima has extended Palmer´s idea. He speaks of situations instead of representing and represented worlds. A situation is charaterized by certain states of affairs holding in it. Shimojima´s claims are based upon the notion of a constraint. Leaving technical issues aside, a constraint governs the way in which states of affairs can hold in a situation. What does this mean with regard to representations? Consider two sets of affairs, SoA1 and

SoA2. If SoA1 cannot hold in a representation without at least one element of SoA2 holding in it, too, then there is a constraint SoA1→ SoA2. Shimojima then introduces the idea of constraint projection. Say we have a representing situation SR and a represented situation SO. Let SR be charaterized by a number of states of affairs, call them $\sigma 1, ..., \sigma n$, each of which represents a state of affair in SO. Due to constraints, the representing vehicle can only represent $\sigma i, ..., \sigma n$ if it also presents some other states of affairs $\sigma n+1, ..., \sigma n+m$. These in turn again represent states of affairs in the represented situation. So the constraint that is at work in the representing situation is projected upon the represented situation.

At this point, again a third kind of similarity appears. This time it is a similarity of constraints. The basic idea has been stated by Barwise and Etchemendy (1990) as follows: "Diagrams are physical situations. They must be, since we can see them. As such, they obey their own set of constraints ...By choosing a representational scheme appropriately, so that the constraints on the diagrams have a good match with the constraints on the described situation, the diagram can generate a lot of information that the user never need infer. Rather, the user can simply read off facts from the diagram as needed. This situation is in stark contrast to sentential inference, where even the most trivial consequence needs to be inferred" (p. 22). Shimojima (2001) explains the difference with the existence of two different sets of constraints. There are two kinds of such constraints relative to the metaphysical status of origin they have. If a constraint is "due to natural laws, such as topological, geometrical, and physical laws" (p. 20), Shimojima labels it a nomic constraint. If it is due to "stipulative rules[...], such as syntactic well-formedness conditions" (p. 20) (conventions governing forms of representation), it is called a stipulative constraint. Shimojima's main claim is that some representations are governed by nomic constraints (plus, possibly, by stipulative constrints) while other are not. Moreover, the existence of nomic constraints is the principal source of the effectiveness in some representational systems. More specifically, Shimojima claims that the difference between graphical and linguistic representations lies in the fact that the former are governed by nomic constraints while the latter are not. According to him it is the mechanism of nomic constraint projection that is responsible for the cognitive power of diagrams. At this point it gets really interesting for our study of diagrams. We could say that a representation is more or less "diagrammatical" depending on the number and complexity of nomic constraints at work in it. Note that the notion of nomic constraints does not entail any specific claims about the medium of the representation. Many physical systems can have a good constraint matching with the representing situation.[8]

The lesson to be learnt here is that diagrams exhibit interesting properties once we start looking beyond the undifferentiated structural similarity claim. Two representations can both be two-dimensional graphics and can both stand in relation of a structural similarity to their objects and still differ greatly in the role which they play in reasoning processes because on of them employs more nomic constraints than the other (if any).

Nomic constraint projection is an influential factor with regard to the role of representations in reasoning processes. Shimojima makes this clear when he identifies and explains often reported key properties of diagrams by appealing to nomic constraint projection. I will briefly discuss the two most important ones: the ability of diagrams to generate free rides, and the unavoidability of content speficity.

1. Free Rides. In diagrams it is often the case that when you present a set of information you will get some further information without having to do anything for it. In some cases the additional information logically follows from your original set of information. You get it for free. That is why it is called a free ride. It is as if the spatial structure of the medium computed the solution for you. An illuminating example is syllogistic reasoning with Euler diagrams. Let us assume you have two pieces of information: "All A are B" and "No C are B". You can easily conclude that "No C are A". Let us now see what happens if we present the pieces of information in the system of Euler diagrams according to the rules set by the convention of how to draw Euler diagrams. If you have never done it before, try it, that helps understanding the idea. First draw a circle labelled A, then draw a circle labelled B around the circle A such that the two circles do not overlap. This represets "All A are B", the first piece of information. Let us now represent the second piece. For this you have to draw another circle, labelled C, that must not be drawn inside but instead has to be drawn outside the circle B. At this point you have represented all the pieces of information you got in the diagram. What is left to do? The answer is: nothing. At this moment the diagram already presents the piece of information that the circle C and the circle A are completely separate (which implies that it C is not included by A). This is because there holds a constraint on the representing situation which has a form like: "There are three circles x, y, z such that they do not

---

[8] Linguistic representations can also be governed by nomic constraints. Say the words "veni", "vidi", and "vici" represent different events in the world and the relation "is left of" represents the corresponding relation "happens earlier than". Then the sentence "veni, vidi, vidi" is governed by natural constraints due to the linear ordering of written language.

overlap and that x spatially includes y and that z is not included by y → z is not included by x". The constraint holds because of topological laws and it is therefore nomic. It is projected via semantical conventions on the level of the represented situation which allows us to simply "read off" the logical inference that "No C are A" from the diagram. Of course standard logical reasoning would have yielded the same result. The important point is not that you get a result only from diagrammatic reasoning but that you get it without having to do something for it. The difference lies in the computational effort that the reasoning subject has to undertake. In the case of this example the difference does not seem to be much. But when dealing with much more complex representations the difference can be huge. A good example for this is the process of information upgrading in maps. Shimojima writes: "Geographical maps have the same function [of free rides, J. W.], and to a much larger extent. Adding an icon of a house to a particular position in a map results in the expression of various new pieces of information, concerning the spatial relationship of the house to many other objects already mapped. This is due to spatial constraints governing map symbols, which are pretty isomorphic to spatial constraints governing mapped objects" (pp. 75-76).

2. Content specificity (also: over-specificity). This feature refers to the fact that you cannot present certain pieces of information without also presenting other pieces of information that do not logically follow from the first set of information. The most notorious example is probably the inability to draw an abstract triangle, e. g., an isocecles triangle, without at the same time giving its sides a specific length. The unavoidability of presenting accidental feature that are not warranted by the set of states of affairs that was necessary to construct the presentation. The property of content specifty holds in a certain representation system whenever you have to present additional information that is not a logical consequence of the set of informations that you used to construct the representation in the first place.

7. The medium of representation: spatial structures

Let us now turn to the kinds of criteria that are used to define diagrams. The claim concerns the medium of a diagram. It proposes to view a diagram as a representing vehicle with the following properties. (i) its relevant inscriptions operate on a two-dimensional surface and (ii) the inscriptions are interpretable as a spatial structure (be it a metrical, topological or any other kind of a spatial structure). The importance of the spatial constitution of diagrams has repeatedly been pointed by scholars. Krämer (2009) considers interspatiality (Zwischenräumlichkeit) to be the constituting property of a wide range of notational forms, ranging from writing to diagrams. Heßler and Mersch (2009) make an interesting distinction between two forms in which an image can be said to be spatial: as in extensum or as in spatium. The first sense refers to the trivial fact that an entity can be extended in space. The concept of spatium, however, refers to "topological differentiality" (p. 27, my translation). What the authors seem to have in mind with this notion is the potentiality of a graphical medium to realize or exemplify a topological structure. Heßler and Mersch take this as the lead principle of any science of diagrams: "Diagrammatically and graphically structured spaces are based on a spatial logic" (p.33, my translation).

This idea gets interesting when we combine it with the idea of intrinsic or nomic constraints of certain representations. Stenning and Lemon (2001) note in this respect: "Diagrammatic representations often exploit non-trivial spatial structure in representation. The price they pay is that they must obey the mereological, topological, and geometrical constraints of the plane" (p. 33). Shimojima and Katagiri (2008) point to the potential benefits of diagrams having to obey spatial constraints. "Many, perhaps all, systems of diagrams have the function of letting the user to exploit spatial constraints on their graphical structure, and thus lightening the load of inferences" (p.74).

We can now better understand the role of the spatial structure of diagrams. The spatial structure carries or even generates information. The important point in the case of intrinsic and nomic constraints is that the structure inherently performs cognitive tasks, from representing a structure to inferring new information, that in other cases the user has to perform himself. [9]

8. Diagrams as cases of cognitive extension

One important aspect about the effect of nomic constraints on the information displayed in a representation is that it turns diagrams into a case of extended mind. This refers to the idea that cognition (at least in many cases) does not take place in the head (or the body) of a human agent alone but involes processes and entitites that are located outside the skin. In their seminal paper "The extended mind" Clark and Chalmers (1998) suggest a

---

[9] Note however, that phenomena like over-specificity show that there is a negative side to this, too. In many cases there is a trade-off between effective reasoning and expressive richness.

parity principle as main criterion for identifiying instances of an extended mind: "If, as we confront some task, a part of the world functions as a process which, were it to go on in the head, we would have no hesitation in accepting as part of the cognitive process, then that part of the world is (for that time) part of the cognitive process" (p. 8). In this perspective a diagram performing a free ride clearly presents a case of extended mind: the spatial structure performs the same function as a mental inference. The satisfaction of the criterion is less obvious in the case of content speficity but it takes place nevertheless. The difference is that here the coginitve process leads to new pieces of information that are unwarranted by the original set. But this could also happen were the process "to go on in the head".

Attributing cognitive performance to diagrams implies an uncommon picture of the role that the spatial structure plays. It does more than merely record information. Instead it plays an active role in the reasoning process. It performs a cognitive task by bearing or generating informational content. Reasoning with diagrams can thus be regarded as a kind of division of cognitive labour between human user and diagrammatic representation. This seems to be a main factor in explaining the often noted effectiveness of diagrammatic reasoning. According to Barwise and Shimojima (1996) "most advantages and disadvantages of surrogate reasoning can be explained with reference to the ways in which the default constraints on surrogates intervene in the process of problem solving" (p. 19). They even claim that "every practice of surrogate reasoning involves a constraint projection [...], and a part of its inferential burden is taken over by a constraint" (p. 20).

9. Defining diagrams

Let us turn to the question of how to define a diagram. I will not present a single characterization. Instead I favor a taxonomy that distinguishes between various different ways of defining diagrams. It is based on the distinction of material and functional properties of a representation.[10] Depending on how material and functional criteria are satisfied, we obtain different kinds of diagrammaticity. They are organized around two key distinctions.

First of all, we can differentiate between broad and narrow notions of diagrams. The prototype of a broad notion is Peirce's operational iconicity

---

[10] This is already implicit in the two claims discussed in the introduction: a diagram consists in (i) the spatially structured graphical medium and (ii) a structural similarity to its object.

criterion which makes claims about the relation between diagram and object but not about the medium of the diagram. It is medium-insensitive. Narrow definitions, however, restrict the medium of the representing vehicle to a two-dimensional graphical structure. One could perhaps also describe the distinction in terms of metaphorical and literal notions of diagrams. Metaphorical notions consider entities as diagrams whenever they stand in a relation of structural or nomic similarity to their object. Literal notions limit the application of the term diagram to graphical structures.

The second distinction concerns the question of nomic similarity. Some proposals for defining diagrams are insensitive to whether there is a matching of nomic constraints between diagram and object. They stop at the level of structural similarity. Other approaches, however, include or even focus on nomic similarity. As we have seen, only the latter approaches enable us to understand certain very important features of diagrams. I would like to describe this difference by calling diagrams that show nomic similarity to their target situation "optimal" diagrams. This labeling is inspired by Peirce's distinction between an "operational" and an "optimal" iconicity that is similar to my proposal (see next section).

10. Peirce

In this section I would like to relate the previous discussion to some of Peirce's ideas on diagrams.

1. In the cases of free rides, nomic constraints force the vehicle to represent more information about a target situation than those that we need to know to produce the representation. When stated like this, the formula is quite similar to Peirce's definition of an icon quoted earlier: "For a great distinguishing property of the icon is that by the direct observation of it other truths concerning its object can be discovered than those which suffice to determine its construction" (CP 2.279). It almost seems as if the definition of free rides was the perfect example for Peirce's operational criterion of iconicity.

2. I have pointed out different ways in which a structural similarity can be established. A diagram whose relations are intrinsically similar to the represented ones can be regarded as "more" similar than a diagram whose relations are only extrinsic. This idea was anticipated by Peirce. It is little known that besides his famous operational criterion Peirce developed

another notion of iconicity.[11] The main idea is revealed in the following passage: "A diagram ought to be as iconic as possible, that is, it should represent relations by visible relations analogous to them" (CP 4.432). Here the notion of diagrams is not only explicitly tied to visual signs but, even more important, the idea of a scale of similarity is introduced. A diagram can be "more or less iconic", depending on how object relations are represented. The more "analogous" the representing relations are to the represented relations the more iconic the sign is. This recalls both Palmer's and Shomjima's analysis. So two signs can be equal in terms of "operational iconicity" but can still differ in terms of "optimal iconicity".

3. The distinctions between broad and narrow notions of diagrams helps to to solve the puzzle with Peirce's operational criterion of diagrams. On the one hand the concept has played a hugely influential role in contemporary research on diagrams. On the other hand many scholars are puzzled by the broadness of Peirce's notion of diagram. In Peirce's writings the term "diagram" refers not only to spatially structured graphical representations but equally to algebraic formulas, writings, etc. In fact, there seems to be little that cannot be used in a diagrammatic way. A good thing about this broadness is that it widens our rather intuitive notions of diagrams and lets us see the diagrammatic in what seems to be not of a diagrammatic nature. But it also seems to leave us incapable of distinguishing graphical representations usually called diagrams from a lot of other representations. This problem, however, can be solved by realizing that Peirce's operational iconicity criterion is insensitive to the medium of the icon.

11. Final remarks

To conclude I will briefly highlight three points that seem to me particularly important.

1. There are many objections to invoking concepts of similarity in discussions of images. Some might also apply to the more limited class of diagrams. Does this render my account flawed? I think not as my aim was not to explain representation by appealing to similarity. Instead, I have taken for granted that diagrams are representations and have then focused on ways in which they are similar to their objects. This is not to say that they derive their ability to represent from this similarity.

---

[11] The following section is based on Stjernfelt (2006).

2. I have distinguished the diagrammatic function of structural (and nomic) similarity from the diagrammatic medium. This distinction helps us to avoid the futile discussion of whether diagrams are intrinsicially graphical or visual or if there can also be non-visual diagrams. The answer to this question depends on whether you are a proponent of a broad or narrow notion of diagrams.[12]

3. At the beginning of this paper I presented the Triple-S-Definition of diagrams. In the course of the investigation we have seen that this definition does not capture an essential property of many diagrams, namely that they stand in a relation of nomic similarity to their objects. To state only the existence of structural similarity is not sufficient because it is precisely this often overlooked third kind of similarity that enables us to identify and explain key properties of diagrams such as free rides. It constitutes the logical basis for the astonishing cognitive power that many diagrams exhibit. With Shimojima one could argue that any diagrammatic representational system that is effective in reasoning processes can be so only because of nomic constraint projection. Most importantly, it is the only form of similarity that actually exploits the spatial structure of diagrammatic mediums. Although it might make sense to label any graphical representation that is structurally similar to its object a diagram, those that involve nomic similarity seem to me the most interesting cases (and the ones one is most likely to encounter in real reasoning processes). Nomic similarity is evidence for the interplay between the spatial structure and the cognitive or logical power of diagrams.

References

Barwise, J., & Etchemendy, J. (1995). Heterogeneous Logic. In J. I. Glasgow, N. H. Narayanan & B. Chandrasekaran (Eds.) Diagrammatic Reasoning: Cognitive and Computational Perspectives (pp. 211–234). Menlo Park, CA: AAAI Press.

Barwise, J.& Hammer, E. (1995). Diagrams and the Concept of Logical System. In J. Barwise & G. Allwein (Eds.) Logical Reasoning with Diagrams (pp. 49–78). Oxford: Oxford UniversityPress.

---

[12] It would be more interesting to ask what properties a medium must have in order to be able to function diagrammaticaly. When does a medium have the potential to serve as a diagrammatic medium?

Barwise, J. & Shimojima, A. (1995). Surrogate Reasoning. Cognitive Studies: Bulletin of the Japanese Cognitive Science Society, 2(4), 7-26.

Clark, A., & Chalmers, D. (1998). The extended mind. Analysis, 58 (1), 7-19.

Gombrich, E. H. (1982). Mirror and Map: Theories of Pictoral Representation. In The Image and the Eye. Further Studies in the psychology of pictorial representation (pp. 172-214). Oxford: Phaidon Press.

Heßler, M., & Mersch, D. (2009). Bildlogik oder Was heißt visuelles Denken? In M. Heßler & D. Mersch (Eds.), Logik des Bildlichen. Zur Kritik der ikonischen Vernunft (pp. 8-62). Bielefeld: Transcript.

Krämer, S. (2009). Operative Bildlichkeit. Von der ‚Grammatologie' zu einer ‚Diagrammtologie'? Überlegungen ünber erkennendes Sehen. In M. Heßler & D. Mersch (Eds.), Logik des Bildlichen. Zur Kritik der ikonischen Vernunft (pp. 94-122). Bielefeld: Transcript.

O'Brien, G. & Opie, J. (2004) Notes towards a structuralist theory of mental representation. In H.Clapin, P.Staines & P.Slezak (Eds.) Representation in Mind: New Approaches to Mental Representation (pp. 1-20), Amsterdam: Elsevier.

Palmer, S. E. (1978). Fundamental Aspects of Cognitive Representation. In E. Rosch & B. B. Llyod (Eds.), Cognition and Categorization (pp. 259-303). Hillsdale, New Jersey: Lawrence Erlbaum Associates.

Peirce, C. S. (1960). Collected Papers, ed. C. Hartshorne & P. Weiss. Cambridge, MA: Belknap Press.

Shimojima, A. (2001). The Graphic-Linguistic Distinction. Artifical Intelligence Review, 15, 5-27.

Shimojima, A., & Katagiri, Y. (2008). An Eye-Tracking Study of Exploitations of Spatial Constraints in Diagrammatic Reasoning. In G. Stapleton,, J. Howse, & J. Lee (Eds.), Diagrams 2008, LNAI 5223 (pp. 74-88). Berlin Heidelberg: Springer.

Stenning, K., & Lemon, O. (2001). Aligning Logical and Psychological Perspectives on Diagrammatic Reasoning. Artifical Intelligence Review, 15, 29-62.

Stjernfelt, F. (2006). Two Iconicity Notions in Peirce's Diagrammatology. In H. Schärfe, P. Hitzler & P.Ohrstrom (Eds.), Conceptual Structures: Inspiration and Application: 14th International Conference on Conceptual Structures, ICCS 2006 (pp. 70-86). Berlin Heidelberg: Springer.

The Extension of the Peircean Diagram Category. Charting the Implications of a Diagrammatical Revolution in Semiotics

by
Frederik Stjernfelt

Different semiotic schools have had different ideas about what counts as the prototypical semiotic phenomenon. In structuralism, the single word relationally defined within the paradigm was probably the prototype, when discussing the sign as such, words like "tree" was taken as the example, and when discussing the relation between related signs, the linguistic paradigm (of e.g. colour terms) was taken as prototype. In Chomskyanism, syntax and recursion were seen as prototypical – the transformation of one and the same syntactic deep structure to different surface linearizations was taken as the basic semiotic phenomenon. In the current wave of cognitive semantics and cognitive linguistics - closer to Peirce's ideas - bodily based image schemata and the mapping of them between semantic domains are taken as the prototypical semiotic phenomena.

Taking, as does Peirce, diagrams as the prototypical semiotic phenomenon invites us to a wholly new way of approaching semiotics. It reconstructs the connections from semiotics to logic and to epistemology and purges semiotics of its tendencies to skepticism and anti-scientific irrationalism so easily taken as the consequence of the idea of signs as being basically arbitrary. The diagrammatical point of view thus takes iconicity to play a very basic role in all kinds of semiotics – but in a way which widely expands the notion of iconicity which is no longer tied to immediate perceptual resemblance only. Iconicity rather covers all sorts of structural isomorphisms between the sign and its object which may not be obvious for a first glance and which may, in many cases, require much effort to establish and expand. This is what lies in what could be called Peirce's "operational" criterion for iconicity:[1] icons are signs from which more can be learnt about the object than what lies explicitly in the construction recipe for the sign. Icons thus contain implicit information – which is why they constitute, in general, the informative aspect of any sign process. This is also the reason why sign uses far from the common sense notion of diagrams - logic,

---

[1] Cf. Stjernfelt 2006

algebra, pictures, linguistic grammar and lingutistic semantics - must be reconceptualized as specific subtypes of diagrams. So Peircean diagrammatology not only develops our understanding of diagrams proper and their importance - it also invites us to redescribe the whole of semiotics and see its connections to logic and scientific knowledge in a new light.

Peirce does not explicitly state that geometric figures are prototypcal diagrams – but it is quite evident that they provide one of the main sources of the general diagram concept in the mature version of his semiotics around the turn of the century (cf the Euclidean wording of his two important subtypes of diagrammatical reasoning, "corollarial" and "theorematical", respectively). The surprising fact that it is possible to prove geometrical theorems by the manipulation of diagrams forms an insight which is basic for Peirce' notion of diagrammatical reasoning. Important aspects of Peirce's notion of diagrams are the following: the typicality of the diagram; its double icon-symbol structure; its facilitating reasoning by diagram experiment; the subtypes corollarial/theorematical reasoning. Let us present these aspects briefly.[2]

The diagram sign vehicle printed or drawn on paper or appearing on a blackboard or computer screen is only a token of the diagram. Several different such tokens may refer to one and the same diagram. Thus, the diagram in itself is a type, and the particular token gives access to the type. A whole series of sophisticated cognitive operations are involved in obtaining this access, idealization, abstraction, generalization – operations which we perform, in many case, without making it explicit. Taking a drawn rectangular triangle as a sign for rectangular triangles as such involve, among others, the following: abstracting from the lines having breadth, abstracting from their not being perfectly linear, abstracting from right angle not being precise, abstracting from the particular colour, size, and orientation of the figure – thus intending an idealized, general rectangular triangle. Furthermore, a thought experiment allows us to imagine the two acute angles of the triangle to vary so as to assume any combination of sizes, thereby letting the particular triangle refer to a whole class of such triangles. Such procedures are involved in the perception and cognition of all diagrams. Thus, the diagram provides direct, perceptual access to an ideal type – this is, to Peirce, a most important aspect because diagrams thus form the cognitive tool for directly accessing idealized, universal states-of-affairs. This is also important for the reason that it removes the diagram from being mental images in the mind of the perceiver only – such mental images are only the means of access to the ideal structure of the diagram as type (and indeed, most external diagrams on paper or screen vastly surpass the

---

[2] For a more thorough presentation, see Stjernfelt 2007, chapters 3-4.

complexity range we are able to grasp by our inner gaze only). The diagram thus provides what is in the Husserlian tradition called "categorial intuition" yielding direct, perception-like access to ideal structures.

Such idealizing reading procedures for the diagram token are tightly connected to the symbolic side of the diagram. Even if basically a schematic icon, the diagram is governed by symbols determining both the specific types of idealization and generalization to be undertaken while reading it – and the possible experiments tried out while using it for reasoning purposes. Thus diagrams explicitly form the inheritor in Peirce's system to Kant's famous schemata from the 1st Critique, forming the bridge between *Anschauung* and *Verstand*, intuition and understanding (corresponding to the iconic and symbolic sides of the diagram, respectively). Thus, in the triangle example above, the operations described are possible only by the explicit or implicit symbolic rules indicated by it being a "rectangular triangle". Without this symbol, the diagram token may be read in other ways (as an example of a triangle in general, as an example of a closed curve, as an example of geometrical figures as such, etc.). Topgraphical maps very directly display this duplicity: the iconic side of the map, displaying the geographical shape of certain landscape features, is supplemented by the symbolic indication of the map's scale, by its legend of symbols used (e.g. a colour scale from green over yellow to brown in order to indicate height, etc.). On top of that, the diagram may use symbolic indices in order to identify its reference object – in the topographic map signs like place names, longitudes and latitudes, etc.

What connects diagrams to epistemology and the acquisition of knowledge, however, is tied to the possibility of making diagram experiments in order to extract implicit knowledge from the diagram. Again, Euclidean figures form the basic prototype: using geometrical diagrams, it is possible to conduct proofs with general validity. In a map, you may judge the distance between two sites on the map by measuring the map distance between the two and divide by the map's scale. You may extract propositions about the relative size of land areas, you may draw conclusions as to the geopolitical position of countries relative to others, and much more. All these propositions are but implicitly present in the diagram but may be made explicit by diagram experimentation. Such diagram experiments cover a very wide span from experiments which are so simple that the result may be read directly off of the diagram and to extremely complicated proofs requiring inventive and ingenious diagram manipulation. This difference partly rests upon particularities in human perception and cognition – human vision, e.g. is geared towards gestalt perception and many gestalt properties in diagrams are easy for human perceivers to grasp immediately, while other such tasks, e.g. metrical area measure, is much more difficult to the human eye. But it is a great insight in Peirce that the difference in diagram experiment difficulty does rely upon human psychology only, but also upon intrinsic diagram properties. This is what is addressed by his distinction between corollarial and theorematic reasoning – he himself proudly called it his "first real discovery" (NEM IV, 49). In corollarial reasoning, the resulting proposition may be read directly off of the diagram. An example could be the diagram of a square with the side s:

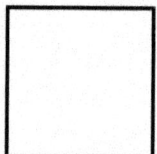

The size of the perimeter of the square may be read directly off of the diagram: as the square has four sides, its perimeter is equal to $4s$. In theorematical reasoning, by contrast, auxiliary entities have to be added to the diagram before the result can be reached. Take for instance the Euclidean proof that the sum of the angles in the triangle equals two right angles. It depends on certain auxiliary lines being drawn:

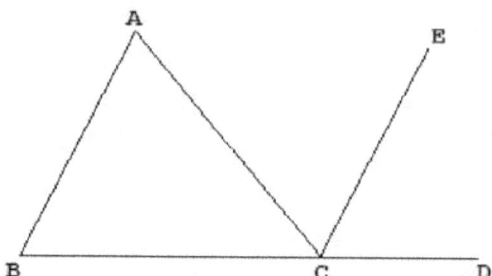

The proof focuses upon the point C. The angle ACE and BAC are equal (because the lines AB and EC are parallel); by the same token the angles ECD and ABC are equal. Then the three angels of the triangle are equal to the three triangles meeting in C: ABC+BAC+ACB = ECD+ACE+ACB. But the sum of the three angles meeting in C evidently equals two right angles.

This proof requires the addition of the two auxiliary lines CE and CD to the original triangle – so unlike the corollarial case, the proof may not be directly read off of the simple triangle diagram. This addition of new auxiliary material to the diagram is what makes the proof theorematical. As Jaakko Hintikka writes, it is Peirce's brilliant insight "... that this geometrical distinction can be generalized to *all deductive reasoning*" (Hintikka 1983, 109). In some cases, we may be able to add the auxiliary entities in our imagination, in other cases, the actual change and manipulation of the external diagram is needed.

The extension of the Peircean diagram category

Given these preliminaries, we are able to appreciate the issue of the extension of Peirce's diagram category. The surprising thing is that Peirce's diagram category extends to such different issues as logic, algebra (and mathematics in general), pictures, linguistics (lexical as well as grammatical issues). This is why the acceptance of Peirce's diagram doctrine calls for a Copernican revolution in semiotics: no general semiotic issue will remain untouched by diagrammatical reasoning. This revolution will be cognitive – because it directs the semiotician's gaze to which cognitive skills are needed in order to process diagrams and acquire knowledge by doing so. This revolution, moreover, is also ontological – because it reemphasizes the connection of semiotics to the issue of ideal structures – mathematical and logical idealities as well as empirical universals. Let us take the single issues one by one.

Logic

The basic insight in the central role of diagrammatical reasoning seemed to have dawned on Peirce in the 1880's when he worked on his "algebra of logic" – which turned out to be the first version of the Russell-Peano formal language later becoming the standard way of expressing modern logic –

> "*The truth, however, appears to be that all deductive reasoning, even simple syllogism, involves an element of observation; namely deduction consists in constructing an icon or diagram the relation of whose parts shall present a complete analogy with those of the parts of the objects of reasoning, of experimenting upon this image in the imagination, and of observing the result so as to discover unnoticed and hidden relations among the parts.*" ("On the Algebra of Logic", 1885, W5, 164; 3.363)

This idea forms the basis of Peirce's doctrine of diagrammatical reasoning: all deductive reasoning is undertaken by means of diagrams. This sweeping claim is connected to the ideality of diagrams: deduction, necessary reasoning, is possible only within domains with absolute truth values – but to a fallibilist like Peirce such truth values only pertain to ideal domains, any empirical domain always being characterized by probable knowledge only. This does not imply that the diagrams may not involve large quantities of empirical facts and data – but diagrams are idealized, general structurings of such data which, accordingly, facilitate deduction. In that sense, the knowledge obtained by diagrammatical reasoning is certain and at the same time hypothetical. It depends on accepting the diagram as a hypothesis and then deriving necessary implications from that hypothesis. Every applied diagram is a proposition – the map of Portugal is a proposition claiming "Portugal has these geographical structures". Thus, a diagram involves a subject and a predicate – an index and a rheme, in Peirce's terminology. The predicative, rhematic side of it is the pure, iconic diagram proper – while the subject is the reference to which object, empirical, ideal, imagined or otherwise, which the diagram predicate is used to portray or describe. Thus, deductive logic as such is diagrammatical. As mentioned, Peirce's 1880s "algebra of logic" is the first version of modern formal logic, and the corollary of deduction as diagrammatical is that such formal languages form prime examples of diagrams. This may come as a surprise, because formal languages are often taken as examples of "symbolic" representations radically *different* from diagrams. It is well known that later in his carreer, in the years around 1900, Peirce constructed a competing set of logic representations in the shape of his so-called Existential Graphs which employ two-dimensional graphs and which may not be linearized like normal formal logic. He did see these representations as "more iconic" than his own linear formal logic systems – but it is important to see that in doing so, he invokes another iconicity concept than the basic operational iconicity concept. For exactly the same set of theorems may be derived from the two competing formalisms (Alpha-Graphs and Beta-Graphs being consistent and complete representations, isomorphic to propositional logic and first order predicate logic, respectively). So a strong implication of Peirce's diagram concept is that linear formal logics are *also* diagrammatical – their iconicity lying in the axioms and deduction rules facilitating which diagram experiments may be made using them when conducting a proof in them.

As logic inferences naturally occur in ordinary speech immediately entails that diagrams must be involved in aspects of language structure as well – we shall return to this issue below.

Mathematics

In "On the Algebra of Logic", Peirce claims about algebra in general that

> "*As for algebra, the very idea of the art is that it presents formulæ which can be manipulated, and that by observing the effects of such manipulation we find properties not to be otherwise discerned. In such manipulation, we are guided by previous discoveries which are embodied in general formulæ. These are patterns which we have the right to imitate in our procedure, and are the **icons par excellence** of algebra.*" (CP 3.363)

The iconic aspects of algebra are thus expressed in the general formulas for manipulation of the single signs – which, as Peirce goes on to say, are symbols only. In an algebraically expressed arithmetical equation like

$$ax + by = c$$

each of the single signs is a symbol. But the rules governing such an equation are given by the algebraic ring of the two operations addition and multiplication which is what allows us to manipulate it and solve it in one of the unknown variables:

$$x = (c-by)/a$$

Again, it is the operational iconicity criterion which establishes the iconicity – and, hence diagrammaticity – of algebraic expressions. The fact that they may be manipulated in order to make implicit properties explicit is thus the crucial feature granting they are diagrams in Peirce's sense. Algebraical diagrams display the enormous span of such diagrams from the most corollarial ($x+2=4$) and to the most theorematical ($x^n + y^n = z^n$ has no integer solutions for $n > 2$) – the latter, the so-called Fermat's last theorem remained unsolved for more than 300 years and took the collected efforts over centuries of many mathematicians and finally many years of computer work by Andrew Wiles to be proved in 1993.

The connection between mathematics and Peirce's claim that all deductions are performed by diagrams lies in his mathematics definition which he inherited from his father, the mathematician Benjamin Peirce: mathematics is the science that draws necessary conclusions – with the addition that these conclusions are drawn from hypotheses. Thus, mathematics becomes the diagrammatical science par excellence, and for a Peircean point of view, diagrams of all sorts and all empirical varieties contain a formal, mathematical kernel.

Linguistics

The diagrammatical reinterpretation of language is an especially complicated task. The overall claim, of course, is that the structuralist doctrine of the arbitrarity of language is incorrect because of taking its prototypical example in the relationship between single words sounds and their meanings.[3] Diagrammatically organized signification seems to be present at many different levels in language. The issue, of course, is which kinds of diagrams pertain to which levels of language.

A basic tendency seems to be that the distinction between grammar and morphology on the one hand, lexical semantics on the other, corresponds to diagrams pertaining to formal and material ontologies, respectively.

Which diagrams pertain to the level of grammar?

When Peirce writes:

> "... if you take at random a half dozen out of the hundred odd logicians who plume themselves upon not belonging to the sect of Formal Logic, and if from this latter sect you take another half dozen at random, you will find that in proportion as the former avoid diagrams, they utilize the syntactical Form of their sentences."
> ("Prolegomena to an Apology of Pragmaticism", 1906, CP 4.544)

Peirce's argument here, of course, is that logic is formal and diagrammatical, even when it is clothed in the garb of ordinary language. The idea here is thus that grammatical forms make logical reasoning possible – which is why such forms must embody diagrams facilitating such reasoning. The immediate linguistic phenomena referred to is, of course, the conjunctions which are formalized as logical functors in propositional logic (and, or, not, therefore, if-then, if and only if, etc.) as well as linguistic quantifiers which may be equally formalized (all, no, some, etc.) – the linguistic structures which makes it possible for ordinary language to involve

---

[3] It should be added, of course, that structuralism did not entirely subscribe to the arbitrarity hypothesis. As Cassirer argued (1945), a central aspect of structuralism is the notion of linguistic wholes (connecting it to field theories of physics and gestalt theories of psychology) – and its charting of such structural wholes in grammar, myth, narratology, may easily be reinterpreted diagrammatically.

in reasoning corresponding to propositional logic and first order predicate logic. Peircean logic, however, does not stop here.

But very basic structures of grammar also pertains to this domain. To Peirce, the basic grammatical distinction on the sentence level is that between rhemes and their saturation by means of indices to form dicisigns – or, translated into linguistics, predicates (verbs, common nouns, and adjectives) and their saturation by means of subjects (represented by nouns, proper names, pronouns etc.) to form sentences. This distinction, in turn, refers to (but is not identical to) the ontological distinction between objects and their relational properties. The main iconic – and hence diagrammatical – part of the sentence is hence the predicate. But this takes us into lexical semantics. Thus, a basic distinction goes between the diagrammaticity of the sentence structure as such and the diagrammaticity of the rhemes – predicates - involved in it. If we remain on the grammatical level, we may make the following observations

1) grammatical transformations have logical content– the active-passive transformation allows for the inference that if Peter beats Paul, Paul is beaten by Peter.

2) grammar and morphology (distinctions between word classes) mostly concern formal ontological properties (while lexical structures generally account for regional ontological properties):

(thus, the morphological distinction between *nouns – adjectives* will correspond to the formal ontological distinction *objects – properties*; *verbs – nouns* to *processes – objects*; *tempus* to *time categorized from the point of view of the present*; *numerus* to *simple number*; *adverb*s to *process properties*; *aspect* to *parts and wholes of processes*; *prepositions* to *relations*. Grammar may also concern a few (but still very general) material ontological issues: persons – other objects (pronouns), force dynamics (modalities).

Thus, as maintained by cognitive linguistics, there is no sharp distiction between grammar and semantics which may be mapped directly onto the sharp distinction between formal and regional ontology – but still, the overall tendency seems to remain that grammar provides a formal ontological framework with rather few material ontological additions – while detail of material ontology is filled in by means of lexical semantics of single word roots of predicates: adjectives, common nouns, and verbs.

These ideas are, of course, related to Husserl's distinction between categorematica and syncategorematica inherited from the scholastics and they have further correlations in Talmy's and others' related distinction between open and closed word classes such that syncategorematica form closed classes (pronouns, conjunctions, inflection morphemes, prepositions, modal verbs), establishing in the sentence a general, topological framework which is specified in the single case by the lexical semantics of the single roots of the categorematical open classes (noun roots, adjective roots, most verb roos).

The picture is further complicated by the fact that subclasses of formal ontological concepts as well as many high-level material ontological concepts also permit formalization and the construction of logical calculi to chart them. To the former belong modal logic, temporal logic, higher order logic, to the latter belong epistemic logic, deontic logic, speech act logic, etc.– with linguistic equivalents in modal verbs, in tempus morphemes of verbs, in propositional stance verbs, in speech act verbs, etc. and all their derivatives in other word classes. At the level of lexical semantics, the cognitive semantics tradition has investigated in great detail the reliance upon schematic – that is, diagrammatical - representations of the semantics of single words as well as their combination in metaphors, blendings, etc.

The general upshot of these sketchy deliberations, however, is that conceived from a diagrammatical point of view, language has two levels, one general, formal and vague, formalized in grammar and closed-class syncategorematica – another level in lexical semantics and open-class categorematica. The former level yield general diagrams on the grammatical and morphological levels; the latter adds material specification by adding more particular diagrams yielding schematic representations of particular kinds of objects, properties and processes in the shape of particular nouns, adjectives and verbs.[4] It is this duplicity which allows for language to be universal in the sense that it – unlike other diagrams – may speak about everything, because widely different material ontological stuff may be inserted in the formal ontological general diagram framework of grammar. This duplicity forms the reason why language is able to talk about any possible subject. This is also the reason why ordinary language – despite many claims to the contrary - is not tied to any specific metaphysics. It is true that distinctions at the grammatical level are not without any metaphysical implications, and it is true that the carving up of the world in linguistic paradigms at the level of lexical semantics may reflect aspects of particular world views – but the separation of these two levels allows for language to reflect upon its own distinctions, modify them, deny them, and thus to house widely different world views. This is also the case because language is *not* universal in another sense of the word: it does not provide

---

[4] See, e.g., the analysis of the adjective "safe" in Bundgaard, Østergaard, and Stjernfelt 2006

the only semiotic access to the world. Different types of diagrams proper – in the shape of diagram tokens of mental imagery, on externally represented on a page or on a screen – constantly interfere with language and may correct language distinctions or claims, just like the opposite may be the case.

This follows from the crucial observtion by Hintikka tied to his distinction between two strands in 20 C philosophy, that of language as a universal representation, and that of language as calculus. The former claims language forms one closed representation system – and from this idea follows the idea that the limits of language forms the limit of the world for the users of that language, the various prisonhouse-of-language hypotheses. A further corollary of this claim is the ineffability of semantics – because the only possibility of addressing semantics is in the same language, which leads to circularity. This leads, again, to the impossibility of defining or discussing truth, because any reference to extralinguistic reality is also caught up within language. Interestingly, Hintikka finds this tradition cuts across the analytical-continental divide in philosophy, thus including not only Frege, Russell, Wittgenstein and Quine, but also Heidegger and Derrida. The other tradition, then, involves Boole, Peirce, Schröder, Hilbert, Husserl, the later Carnap – and Hintikka himself. This tradition sees logic and language not as one closed system, but as a multiplicity of different representation systems with different generality, different aims and techniques. Maybe surprisingly, this representational pluralism, unlike linguistic universalism, is compatible with realism: different diagrams, formalizations, and linguistic representations of the same object are possible, thus semiotically triangulating that object. Furthermore, the semantics and truth claims of one representation systems may be critically assessed by another. This refutes the ineffability claim about semantics, making semantics the possible object of scientific study, and making truth claims of one representation the possible object of investigation by others. The diagrammatical point of view would add to this picture that language does not even constitute one such system. Ordinary languages not only embed highly different logics; their basic distinction and very loose coupling between grammar and lexical semantics on the sentence level as well as their trans-phrastic combination into different genres provides language with an internal plurality which endows it with a plasticity already removing it from the simplistic language-as-universal-representation view. This is evident from the fact that grammatical distinctions may, in many cases, be expressed also by semantical means.[5]

---

[5] French and many other Roman languages have a future tense verb inflection; Germanic languages like English and Danish do not, but represent future by means of a semantic paraphrase using the verb "will".

What one language expresses by means of grammar, another language may express by means of other semantic constructions – and one and the same language may have several, competing ways of expressing the same thing. Thus, the distinction between the two is not sharp and is subjected to a sort of rivalism even within one language.

The isomorphism between basic structures in formal (and material) ontology, in formal logic, and in grammar/morphology, respectively is a very important issue: analogous structures seem to appear in ontology (object-relation), logic (subject-predicate), morphology (noun-verb) and grammar (noun phrase-verb phrase). This does not imply, however, that the four levels refer to one and the same thing. It is not the case that logical formulae nor grammatical expressions necessarily refer to ontological objects. Because of the device of nominalization, properties, relations and other ontological phenomena may make their appearance as subject nouns ("redness", "distance", etc.) and thus occur as logical subjects and grammatical noun phrases. Similarly, the logical subject of an argument need not appear as the grammatical noun phrase in its corresponding linguistic representation. ("Socrates is mortal" – "Mortality is a property rightly ascribed to Socrates"). Thus, this duplicity of isomorphism and independence between ontology, logic, and language contributes to the plasticity of language: it allows for it to investigate higher-order-objects and their behaviour, to articulate counterfactual, contradictory, ontologically unsound claims or claims confounding different material ontologies. This implies language is very far from forming a *mathesis universalis* – even in true claims we should not expect noun phrases to refer directly to first-order ontological objects in the world. Quite on the contrary, this is what facilitates language to be a *calculus ratiocinator* – language may be used to investigate the diagrammatical implications of very different hypotheses, fine-tuning linguistic representation to expressing even very strange and counterintuitive claims.[6] Thus, this allows for language to speak about objects and their properties without 1) logical consistency ("the round square") and 2) without ontological commitment (breaking material laws: "talking stones" - merely physical objects being alive and intending; mixing different (onto-) logical levels: "green virtue"). This makes language an experimental tool where different such combinations may be investigated in a trial-and-error

---

[6] Diagrammatical reasoning allowing for two participants taking turns in manipulating one and the same diagram also makes such reasoning the object for a pragmatic, intersubjective approch to semiotics. Pietarinen (2006) points to the fact that Peirce's diagram graphs involve such a dialogical structure with two persons playing a game trying to prove, viz. disprove a given claim. With respect to the realization of such games in linguistics, this points to the interconnection of semantics and pragmatics.

perspective. This, again, fits the idea of "fallibilistic apriorism" – if it is indeed the case the apriori structures of different material domains are far from known in detail and, in fact, form part of the goal of the ongoing scientific investigation, we should expect language to be able to express many different, conflicting accounts of those ontologies and furnish the possibility for experimenting diagrammatically upon such expressions. By contrast, if Kantian apriorism was correct, language would simply be unable to express propositions breaking apriori laws, because all our intutions and understandings necessarily would conform to such laws, the only open issue remaining would be their composition in schemata. In fallibilistic apriorism, such schemata and their experimental combination in language rather forms the royal road to the investigation of such structures.

A Peircean diagrammatological revolution in semiotics will involve the arduous task of reanalyzing language in terms of diagrams and diagrammatical reasoning. The cognitive semantics and cognitive linguistics tradition during the last 30 years – Fillmore, Lakoff, Johnson, Turner, Talmy, Fauconnier, etc. – provides many detail analyses of particular linguistic structures in terms of mappings and variations of schemata which may be integrated in such a diagrammatological linguistics.

Conclusion

The extension of the Peircean diagram concept in semiotics thus covers all subfields of semiotics where reasoning based on schematic representations occur. I have here made the case that logic, mathematics, and language form central examples of such domains. Other such domains include pictures, cognition, and scientific investigation which also involve schemata endowed with experiment and thus reasoning possibilities[7]. Peircean diagrams thus form an actual and exciting program for semiotics. It makes possible the unification of otherwise scattered subjects of semiotics, it makes possible the connection of them to the basic motive of cognitive understanding and social communicating of schematic meaning, and it opens a vista of important new and deep research questions – such as the task of constructing a rational taxonomy of diagram subtypes.

---

[7] I have given some arguments for diagram use in picture analysis and in interpretation seen as an investigation procedure in Stjernfelt 2007.

References

Peer Bundgaard 2004 "The Ideal Scaffolding of Language: Husserl's fourth *Logical Investigation* in the light of cognitive linguistics", in *Phenomenology and the Cognitive Sciences* 3: 49-80.

Peer Bundgaard, Svend Østergaard, and Frederik Stjernfelt 2006 "Waterproof Fire Stations. Conceptual schemata and cognitive operations involved in compound constructions", in *Semiotica*, vol. 161 – 1/4, 362-393.

Ernst Cassirer 1944 "The Concept of Group and the Theory of Perception", in *Philosophy and Phenomenological Research* 5: 1-35 (original French version 1938).

—— 1945 "Structuralism in Modern Linguistics" in *Word*, vol. I, nr. II, N.Y. 1945.

—— 1954 *Philosophie der symbolischen Formen* I-III, Darmstadt 1954 (1923-29).

—— 1956 "Zur Logik der Symbolbegriffs", in *Wesen und Wirkung des Symbolbegriffs*, Darmstadt (1938).

—— 1985 "Die Sprache und die Aufbau des Gegenstandswelts", in *Symbol, Technik, Sprache*, Hamburg (1927).

Fauconnier, Gilles, and Turner, Mark 2002 *The Way We Think. Conceptual Blending and the Mind's Hidden Complexities*, N.Y.: Basic Books.

Husserl, Edmund 1970 *Logical Investigations* (transl. J.N.Findlay), London and Henley: Routledge and Kegan Paul.

—— 1971 *Ideen zu einer reinen Phänomenologie und Phänomenologischen Philosophie*, drittes Buch *Husserliana* vol. V Tübingen: Max Niemeyer.

—— 1973a *Experience and Judgment*, London: Routledge and Kegan Paul.

—— 1975 *Logische Untersuchungen* I, *Hua* XVIII, Den Haag: Martinus Nijhoff, 1975.

—— 1980 *Phantasie, Bildbewusstsein, Erinnerung*, *Hua* XXIII, Dordrecht etc.: Martinus Nijhoff 1980.

—— 1980a *Ideen zu einer reinen Phänomenologie und Phänomenologischen Philosophie*, Tübingen: Max Niemeyer (1913).

—— 1984 *Logische Untersuchungen*|II, I.-II. Teil (Text nach *Hua* XIX/1-2), Hamburg: Felix Meiner 1984.

—— 1985 *Erfahrung und Urteil*, Hamburg: Felix Meiner 1985 (1939).

—— 1991 *Ideen zu einer reinen Phänomenologie und Phänomenologischen Philosophie*, zweites Buch *Husserliana* vol. IV Tübingen: Max Niemeyer.

Lakoff, George 1987 *Women, Fire, and Dangerous Things*, Chicago: Chicago U.P.

Lakoff, George and Mark Turner 1989 *More Than Cool Reason. A Field Guide to Poetic Metaphor*, Chicago: University of Chicago Press.

Lakoff, George and Johnson, Mark 1999 *Philosophy in the Flesh*, Chicago: University of Chicago Press.

Lakoff, George and Rafael Nuñez 2001 *Where Mathematics Comes From. How the Embodied Mind Brings Mathematics into Being* New York: Basic Books.

Peirce, Charles S. 1976 *New Elements of Mathematics* [NEM], (ed. C. Eisele) I-IV, The Hague: Mouton.

—— 1992 *Reasoning and the Logic of Things*, (eds.Ketner og Putnam), Camb.Mass

—— 1992 *The Essential Peirce,* vol. I. (1867-1893) (eds. N. Houser and C. Kloesel), Bloomington: Indiana University Press.

—— 1997 *Pragmatism as a Principle* [PP], (ed. A. Turrisi), Albany: SUNY Press.

—— 1998 *The Essential Peirce*, vol. II (1893-1913) (eds. N. Houser and C. Kloesel), Bloomington: Indiana University Press.

—— 1998 *Collected Papers* [CP, references given by volume number and paragraph], I-VIII, (ed. Hartshorne and Weiss; Burks) London: Thoemmes Press 1998 (1931-58).

——"Logic, Considered as Semeiotic", constructed from manuscript L 75 by Joseph Ransdell (http://members.door.net/arisbe/menu/library/bycsp/l75/ver1/l75v1-01.htm).

Petitot, Jean 1985 *Morphogenèse du sens*, Paris: PUF.

—— 1992 *Physique du sens*, Paris: Éditions du CNRS.

Pietarinen, Ahti-Veikko 2006 *Signs of Logic. Peircean Themes on the Philosophy of Language, Games, and Communication*, Dordrecht: Springer.

Poli, Roberto „Descriptive, Formal, and Formalized Ontologies", in D. Fisette (ed.) *Husserl's Logical Investigations Reconsidered.* Dordrecht:Springer,183-210 (http://www.formalontology.it/essays/descriptive-ontologies.pdf).

Smith, Barry 1987 "Logic and Formal Ontology" J. N. Mohanty and W. McKenna, eds., *Husserl's Phenomenology: A Textbook*, Lanham: University Press of America (1989), 29-67.

—— 1992 "An Essay on Material Necessity", in P.Hanson and B. Hunter (eds.) *Return of the A Priori* (*Canadian Journal of Philosophy*, Supplementary Vol. 18).

—— 1994 *Austrian Philosophy,* Chicago: Open Court 1994.

—— 1996b "In Defense of Extreme (Fallibilistic) Apriorism", in *Journal of Libertarian Studies,* 12, 179-192.

Stjernfelt, Frederik 2006 "Two Iconicity Notions in Peirce's Diagrammatology", in *Conceptual Structures: Inspiration and Application. Lecture Notes in Artificial Intelligence* 4068, Berlin: Springer Verlag, 70-86.

—— 2007 *Diagrammatology. An Investigation on the Borderlines of Phenomenology, Ontology, and Semiotics*, Dordrecht etc.: Springer Verlag.

Talmy, Leonard 2000 *Toward a Cognitive Semantics*, I-II, Camb. Mass: MIT Press.

Turner, Mark 1996 *The Literary Mind*, N.Y.: Oxford U.P.

Is Non-visual Diagrammatic Logic Possible?

by
Ahti-Veikko Pietarinen

"We can hardly but suppose that those sense-qualities that we now experience, colors, odors, sounds, feelings of every description, loves, griefs, surprise, are but the relics of an ancient ruined continuum of qualities, like a few columns standing here and there in testimony that here some old-world forum with its basilica and temples had once made a magnificent ensemble." (CP 6.197)

Abstract

This note addresses the question of the possibility of logic that has no visual and no written appearance: no symbols, no marks, no language. Is such logic conceivable at all? I will sketch a positive answer that builds upon the idea of diagrammatic logic suggested by Charles Peirce. According to Peirce, the category of diagrams exceeds the visual forms of representation. My case study here concerns developing a propositional logic based on sounds.

Non-visual Representations

Contemporary society is perfused with signs and representations that are predominantly visual. Most of the stimuli seem to reach our consciousness through our ophthalmic pathways. Yet non-visual representations have always been just as ubiquitous. A concerto, a tender bodily touch or a fiery nip of Bhut Jolokia all evoke representations having at least an equal status to those of musical sheets, love letters, or innocent images in a cookbook. Combinations of different modes of representation are therefore routinely affixed with the term 'multimodal'. An increasing number of artifacts and contemporary works of art fall within that category, for instance.

My interest here nonetheless lies not in the broad question of the nature and modes of representation in the arts, but in the question of the possibility of non-visual representations of complex, intellectual propositions. I will likewise leave aside the complex philosophical discussion on the possibility of music to express, represent or refer to something external to it. By non-visual representations of complex propositions I mean the possibility of construing a language-like system of signs the behaviour of which would largely agree with those of written or spoken languages, be they the artificial and formal languages of logic or the natural languages of linguistic communities. There are pictorial languages that can represent facts and propositions (Bogdan 2002, Neurath 1936), so why could not there be non-pictorial and non-visual systems of signs that can do the same?

Charles Peirce devoted a great deal of his life to the investigation of diagrams. He believed diagrams are the proper home for the 'language of thought'---pardon the misuse of this deceptive anachronism---which submit the essential features of the assertions of facts to the control of exact reasoning:

We form in the imagination some sort of diagrammatic, that is, iconic, representation of the facts, as skeletonized as possible. The impression of the present writer is that with ordinary persons this is always a visual image, or mixed visual and muscular; but this is an opinion not founded on any systematic examination. (CP 2.778, 1901, Dictionary of Philosophy and Psychology)

Beginning with the category of diagrams, Peirce developed an exact and rigorous logical theory of how to diagrammatically represent complex assertions and ideas, known as the theory of Existential Graphs (Pietarinen 2006).

Stjernfelt (2007), among others, has investigated Peirce's general notion of diagrams, and concluded that it is wider than the logical notion of diagrams. This question certainly deserves much further study (see Pietarinen & Stjernfelt 2009 for a starter), but my finger now likes to point at Peirce's remark above concerning diagrammatic representations which are visual images "with ordinary persons". As this claim is "an opinion not founded on any systematic examination", there seems to be a lot to be said about the possibilities of non-visual diagrammatic representations.

No tolerably systematic examination can be claimed in this paper, either, and so my strategy is confined to providing one suggestion as to how we could go on to construe such diagrammatic representations which are comparable in their expressive power with that of propositional logic but which have next to nothing to do with visual images. It is in fact surprising that no one has proceeded to suggest anything comparable to this idea, which we may call the 'logic of sounds'.

Auditory and Visual Diagrams

The basic idea is very simple. We take seriously Peirce's suggestion in his 1892 paper The Critic of Arguments that a "diagram has got to be either auditory or visual, the parts being separated in the one case in time, in the other in space" (CP 3.418). Here Peirce stretches the concept of a diagram to accommodate various auditory signals, sounds and hearable stimuli. We have been accustomed to think of diagrams as analogous to geometrical figures, visually perceptible images consisting of an arrangement of marks or drawings such as lines, points, angles and dimensions taking place in some suitable manifold of space and dimension. However, if diagrams may really be both visual and auditory, a lot remains to be studied as regards to the fundamental nature of the latter.

That a diagram can consist of parts separated in time is an intriguing proposal. Peirce never went on to follow up on this line of thought in detail, but he noted that diagrams are in general formed by algebraic methods, and that algebra gives rise to both visually perceptible geometric diagrams and auditory diagrams produced by speech:

> "[A] method of forming a diagram is called algebra. All speech is but such an algebra, the repeated signs being the words, which have relations by virtue of the meanings associated with them." (CP 3.418, 1892, The Critic of Arguments)

Peirce's assertion that all speech is a diagram-forming algebra needs some explication. An algebra is a tuple consisting of sets of elements and operations among them. In this light, a speech is an algebra of lexemes with meaning as the basic set and the operations on that set construe the linear arrangement of the elements of this basic set. He mentions "meanings", which makes algebra not a method for the study of syntax or the formation of grammatically acceptable units of language; rather, it suggests that we need to ascribe to the "syntax follows semantics" thesis: what is grammatically acceptable must be determined on the basis of what the expressions mean, something that enables us to construct a meaningful system of speech and language in the first place.

All speech is algebraic in nature and all algebra is diagrammatic. Since speech is sounds built up from simple phonemes, the diagrams it creates are auditory rather than visual. The 'algebra of speech' is a linear structure, maybe a ring. Of course, there may be other kinds of diagrams taking place in temporal rather than spatial manifolds which are created by sounds and which are likely to be more complex than those created by human speech. In fact, human speech has some fundamental expressive limitations according to Peirce:

> *"There are countless Objects of consciousness that words cannot express; such as the feelings a symphony inspires or that which is in the soul of a furiously angry man in the presence of his enemy"* (MS 499, 1906, On the System of Existential Graphs...)

We can do better and express such feelings in certain kinds of diagrams, Peirce is convinced here. Feelings are not singular emotions, they are guided by habits that are real generals. When we get really infuriated on something, or really affected by something, such as hearing a universally inspiring piece of classical music, we immediately begin to see all kinds of logical diagrams revolving in our minds.

But could sounds as such suffice as being of the nature of diagrams? Recall that Peirce thinks that in certain cases the connection between an icon and its object is a continuous rather than a discrete one. When we are contemplating an icon, he says, we are often contemplating its object at once: "The diagram is for us the very thing" (CP 3.362). In other words, the icon can be its own object. This is what can happen in model theory of logic, for instance, though the more familiar example Peirce gives is from what happens when we are contemplating a painting.

Suppose then that the diagram is auditory. There is generality in its signification, and there must be an abstraction from any particular cases. When we hear a symphony, we may well "lose the consciousness" that the piece is not "the thing" itself (CP 3.362), so well can the best performances of the best compositions achieve their purpose. The momentary state of "a pure dream" is a feeling with at least some degree of generality and can at least in principle be shared or communicated in terms of diagrammatic abstractions. An auditory abstraction seems to satisfy the fundamental criterion of what it means to be an iconic diagram: it is one that has to reflect continuous connections between "rationally related objects" (MS 293:11).

The Logic of Sounds: A Non-visual, Non-symbolic Logic

To move on to the primary case of logic, my general and I believe commonly accepted assumption is that the representation is a much broader phenomenon than the one using visual and perspectival modes. What was said above points out that what is commonly describes as having 'non-propositional content' is certainly one case in point where representation really refers to a wide class of activities many of which cannot and should not be restricted to visual modes of perception.

Peirce's logic of diagrams (Existential Graphs, EGs) was build on the idea that we can have non-symbolic means of expression for logical constants. These means are, as far as possible, iconic, which means that logical constants must strive to show or communicate their own meaning. These icons may be aided by conventions, but it does not turn them into symbols, as such conventions pertain to the rules of manipulating the constants and the rules of inference (transformation rules). The core meaning of logical constants and the complex expressions build upon those constants remains strictly iconic.

For example, the juxtaposition shows the meaning of 'conjunction' in terms of an arrangement of propositions within the same area of the space of assertions, and the cut shows the meaning of a negation as denial in terms of a severance operation of the interior of the cut from the space of assertions. From these two fundamental logical constants the rest of the more complex propositions can be generated, giving rise to the alpha part of the logic of EGs. In more expressive iconic formalisms such as the beta graphs of EGs, a continuous connection and the attachment to bounded regions of space (predicate terms) denote quantification.

How does the similar idea work with auditory diagrams? In that context, the parts are arranged in the temporal sheet of assertion rather than the temporally static spatial manifold. Corresponding to the atomic propositions of alpha graphs that are simple bounded regions of a space is a simple sound of some definite, restricted-bandwidth frequency. Let us call these atomic elements of the system of sounds tones. They have some distinctive quality, such as a pitch, timbre, loudness, on the basis of which they are taken a note of and identified in the minds of the auditor. Tones correspond to the notion of atomic propositions in sentential logic.

Tones are heard against the backdrop of all possible sounds. That is, if no tone is being played at all, all sounds can be brought within the field of an auditory system. The empty sheet of assertion is therefore an infinite-bandwidth white noise extending indefinitely in time. It means a tautology.

One or more tones can be heard at the same time, and such simultaneous arrangements of groups of tones I will call chords. It is easy to see (or rather to hear) that a chord corresponds to what in symbolic logic is a conjunction of propositions. A more precise inductive definition naturally ascertains that a chord can have other chords as its constituent.

We need a 'cut' still, and its place is taken by a definite sound that will not be heard against the backdrop of the sheet of assertion. Let us call it a mute. It is a sheer theoretical notion, since in practice we would need an access to the information concerning the identity of the tones intended but not actually played out. In music, for example, a pause behaves differently: it only denotes anything that is silent, whereas a mute refers to the particular tone that is not brought to the fore after all, a little like a cross-off of a note or a chord from a composer's score. The mute behaves just as a linguistic negation does, because the mute is thus capable of denying that some tone or a complex of tones can be the case at a particular point in time in the temporal sheet of assertion.

For the system to be classical, a double mute is taken to be the original heard sound again. The mute thus behaves like a switch: what is twice muted will bring back the original. One can well contemplate some intuitionistic version in which the second operation does not bring back the original sound, but we leave aside these variations.

I regret that I cannot give examples of some well-formed auditory graphs of this 'language of sounds' within the confines of this report in a written media, but it should be obvious that from the building blocks of tones (atomic propositions), chords (conjunctions) and mutes (negations) all the rest of the well-formed sounds can be construed. A disjunction of two tones is a mute of a chord consisting of two muted tones, and an implication between two tones is a mute of a chord consisting of a 'heard' tone (the antecedent) and the muted tone (the consequent).

The contradiction is a little curious, since it is a mute taking place on the sheet of assertion devoid of tones, in other words a mute of the infinite-bandwidth white noise: it must therefore be an absolute silence. In a contradictory state of affairs, nothing whatsoever can be heard.

The definitions of tones, chords and muted tones and chords give rise to the well-formed system of the 'language' of the logic of sounds.

Given this sketch of the 'language', then how about the proofs? Let us add some more terminology. We call complex propositions phrases that are composed of chords and mutes. A phrase is thus a combination of chords (chords consist of heard and muted tones) that have extension in finite time. As noted, we call any tone and any phrase that is formed in the manner defined here a sound.

We also call the extensions in the temporal dimension in which any sound can be heard or in which any sound can be muted bars. Moreover, a heard bar is a bar not muted or muted an even number of times. A silent bar is a bar muted an odd number of times. Presented with a heard bar, at least one sound will be resonating. That is, minimally the white noise is being heard. A silent bar blocks any stimuli from the field of the auditory system of the auditor of these sounds.

Now we are ready to give the proof rules. They are akin to the sound and complete set of transformation rules that Peirce came up with in his visual diagrammatic logic of the alpha part of the EGs as early as in the 1890s.

Any sound can be doubly muted, and any doubly muted sound is a sound.

Any sound can be deleted from a heard bar.
Any sound can be added to a silent bar.
Any sound can be copied to/deleted from a subsequent bar.

The first rule ascertains that the behaviour of the mute operation is just what the rule of double negation in propositional logic is. The second is 'downward monotonicity': If I hear a sound, I hear anything contained in it, and thus I can safely remove any of the component sounds. This justifies the soundness of the second rule. A special case is that any tone can be left out from a chord. Take a visual analogue: if there are ten people in the classroom where I am presenting a talk, I can state that there are three people in this classroom without having been committed to uttering a falsehood.

The rule three takes as a starting point a silent bar and states that any sound can be played within it. The justification here is that the result will indeed be that nothing will be heard, since anything played within a silent bar remains silent. It is a kind of 'upward monotonicity' of inference in negative contexts, in which I can safely add to something that readily occurs in contexts that deny its truth. The rule of sound addition is therefore a sound rule.

The fourth, the copy and deletion rule, has to do with the repetitions, and cancellations of repetitions, of sounds. Having performed a sound, no matter whether the bar where it exists is a heard or a muted one, I can always repeat it subsequently. What is once heard can always be heard again in bars succeeding the first occurrence of a sound. When I hear a phrase, I am entitled to infer that the same phrase can be heard again in the future. In addition, I can always duplicate a sound in the same bar at will, as the operation of duplication does not alter anything in the original sound. Conversely, a repetition of sound (muted or not) can be cancelled: what can be heard again can equally be not heard again. When I hear a phrase repetitively, I can infer that I can hear the original occurrence of the phrase without committing to any of its repetitive occurrences (or committing just to some). Likewise, I can always cancel a duplication of a sound within the bar in which the sound is being heard.

I hope these explanations of the behaviour of the inference rules for sounds suffice to convince the reader that inferences based purely on the stimuli of sounds can behave in a surprisingly analogous fashion with the commonplace inferences familiar from the symbolic propositional systems of natural deduction.

Conclusions

One general point of the present exercise has been to encourage philosophers, logicians and cognitive scientists to think of logical systems and the basic notions of inference in considerably wider terms than before. That is, logic is not and need not be bound to ordinary symbolic modes of representation per se. And logic may even not be bound to some other kinds of visual yet non-symbolic modes of representations, such as logical diagrams written or scribed on some spatial media of the sheets of assertion.

My case study has therefore been to present a logic of sounds building on an auditory reinterpretation of the concept of a diagram. Equally conceivable are other kinds of non-visual diagrammatisations, including those emerging from haptic, tactile and olfactory stimuli and senses. I could, for instance, create a sound and complete system of inference based on the upward and downward entailments, repetitions, and cancellations of smells, at least in principle. It is patent that such 'logics' are far from optimal for human performance, communication and comprehension, whose sensory mechanisms are brutally limited on these fronts. A machine, or another living species whose sensory mechanisms have not been equally degraded, could comprehend these logics much better that the homo does. Humans are seriously handicapped already with respect to the capability of processing auditory stimuli for any logical and inferential purposes, and the reason must be that the dominance of symbolic and visually iconic systems of representation that the human mind has developed has turned them "the relics of an ancient ruined continuum of qualities". Peirce may have himself been a little unalike, given his self-assessment as to how "the connections between different parts of by brain must be different from the usual and presumably the best arrangement; and if so, it would necessarily follow that my thinking should be gauche" (MS 623, 1909).

Several issues have been left tacit. Nothing was said on the possibility of more expressive extensions beyond the simple sentential level. What could the 'sound' of a quantification be, for example? I expect it to do with the overall idea of diagrams exhibiting "continuous rational connections". In the case of auditory diagrams, then, that idea could involve taking into account something like the sustainability of specific kinds of tones crossing the temporal boundaries of bars.

There are likely to be some applications for the systems of multimodal diagrams which can represent basic logical notions without any symbolic languages and notations. And if humans, the 'Symbolic Species', cannot quite make use of those systems, maybe some day someone else will.

Acknowledgments

Supported by the University of Helsinki Research Funds and the Jenny and Antti Wihuri Foundation. My thanks to Alex Gerner for organising the Lisboa Diagrammatology and Diagram Praxis Workshop, to its participants for a number of comments, and to Professor Frederik Stjernfelt in particular for the white noise and some nearly-devastating criticism.

References

Bogdan (2002). The Semiotics of Visual Languages. Boulder: Colombia University Press.

Neurath, Otto (1936). International Picture Language: The First Rules of ISOTYPE, London: Kegan Paul. Reprinted, with a German translation by Marie Neurath, by University of Reading Department of Typography & Graphic Communication, 1980.

Peirce, Charles S. (1931–58). The Collected Papers of Charles Sanders Peirce, 8 vols., edited by Charles Hartshorne, Paul Weiss, and A.W. Burks. Cambridge, Mass.: Harvard University Press.

Peirce, Charles S. (1967). Manuscripts. Houghton Library Microfilms.

Pietarinen, Ahti-Veikko (2006). Signs of Logic: Peircean Themes on the Philosophy of Language, Games, and Communication, Dordrecht: Springer.

Pietarinen, Ahti-Veikko & Sjernfelt, Frederik (2009), to appear.

Stjernfelt, Frederik (2007). Diagrammatology. Dordrecht: Springer.

II. Visual Diagram Praxis

Picture-proofs in Mathematics:
a Chapter in James Robert Brown's Philosophy
of Mathematics, with Examples

by
Augusto J. Franco de Oliveira

Abstract

Pictures and figures accompanying mathematical proofs have been around for many centuries, fulfilling a basic need for visualization and understanding. Their role in proofs has been the subject of some discussion in the literature, namely, as to whether they can at times constitute proofs by themselves. We present a brief sketch of James Robert Brown's view on this matter and on aspects of visual reasoning expounded in a recent book on Philosophy of Mathematics, with some illustrative examples.

Introduction. Some pitfalls

Mathematicians of all times have used pictures, figures or diagrams, mainly but not only in Geometry. They have used figures for many purposes: illustrating theorems, in Geometry and other areas, as a visual aid, for pedagogical reasons, and sometimes they even proposed pictures as proofs (e.g. in classic Chinese style—proof by contemplation—but also in some western civilizations). Euclid's geometry is said to be the geometry of geometric constructions with ruler (straight edge) and compass, but from the very beginning there were problems, although some of these were only recognized many centuries later. For instance, in the very first proposition of Book I of the Elements an equilateral triangle is constructed, given one side.

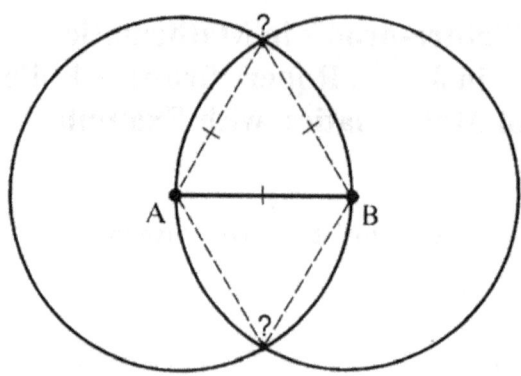

Figure 1 From the proof of Prop. I.1 in Euclid's *Elements*.
The existence of the points where the two circles
meet was justified only late in the 19th century.

We learn in primary school how to make this construction, exactly the same way Euclid did, but no one mentions now, at this level of teaching, and neither did Euclid, that it is necessary to justify that the two circles drawn in fact intersect (in two points). This lack of justification seems to have been first noticed by Leibniz, and is just one of many examples why mathematicians have been and still are suspicious of pictures in proofs. Euclid simply took for granted intuitive continuity properties of lines and circles, as did most if not all mathematicians until the beginning of the 19th century, when B. Bolzano made his first efforts (1817) to turn the intuitively obvious properties of continuous functions into rigorous analytic proofs. Actually, the problem of complete justification of Euclid's Prop. I.1 of the Elements only came about around 1872 with the constructions of the system of real (rational and irrational) numbers ($\mathbb{R}$) made (independently) by R. Dedekind, G. Cantor and K. Weierstrass, and the establishment of the basic continuity properties of Euclidean lines. The figure is not wrong, only its rigorous justification is incomplete.

Dedekind, Cantor, Weierstrass and others, from the seventies of the 19th century onwards, inaugurated the so-called arithmetization or rigorization of Analysis, the main aim of which was to establish the foundations and the development of the infinitesimal calculus in purely analytical terms, independent of the geometric intuition of curves which had been in use since the discovery of modern calculus (Newton and Leibniz). *Hélas*, they unsuspectedly replaced the old and trusted geometric intuition by a still much more problematic one, that of *infinite sets*, but this is another story. Suffice it to say that the (very profound) problems in the foundations of mathematics—the paradoxes that were found in Cantor's intuitive set theory and in Frege's logical system—in the turn of the 19th to the 20th centuries did not arise from the alleged 'lack of rigor' of visual reasoning but from the mathematical principles admitted and verbal/symbolic reasoning. The same can be said of all other great crises in the history of mathematics, like the crisis concerning incommensurable quantities in the time of the ancient Greeks or the crisis concerning infinitesimals (infinitely small numbers and quantities) in the 17th century. More recently, with the development of computers, computer generated graphs (for instance, in fractal geometry), experimental mathematics and computer generated proofs in mathematics the discussion has been revived, but is out of scope of this short paper. For a recent survey with many more examples and relation to the Philosophy of Mathematics see the contributions by Detlefsen and by Borwein in Gold and Simons (2008).

The fear of pictures has some paranoid aspects. Lagrange wrote the *Mécanique Analytique* (1788) without any figures, only algebraic operations were needed, subject to a regular and uniform procedure, he said. Well into the 20th century one of the most influential group of mathematicians, Bourbaki, and some prominent individual members of this group like Jean Dieudonné, urged and put in practice in their works a strict adherence to axiomatic methods, with no appeal whatsoever to "geometric intuition", at least in the formal proofs: a necessity which we have emphasized by deliberately abstaining from introducing any diagram in the book. (Dieudonné, 1969, *Preface ix,* cited by Brown (2008), p. 198).[1]

However, readers of Bourbaki and Dieudonné are supposed to have had a good working knowledge of classical analysis, presumably one with lots of pictures, as found in most if not all university calculus and geometry textbooks. Another prominent member of the Bourbaki group, Pierre Cartier, later commented jokingly: "The Bourbaki were Puritans, and Puritans are strongly opposed to pictorial representations of truths of their faith." (Senechal, 1998, p.28, cited by Brown, 2008, p. 199)

But Brown (2008) goes a bit further in suggesting (Chap. 3) that pictures *in some cases* can serve as perfectly rigorous evidence. More on his views below.

---

[1] Dieudonné even wrote an elementary textbook on linear and analytical geometry without any figure, aiming at reducing the gap between secondary and university education. In the Introduction to Dieudonné (1964), p. 15 he says (my translation) "I also refrained from introducing *any figures* in the text, if only to prove that we can do without them very well; but still this is an omission that my readers can overcome by themselves."

It is true that pictures can be deceiving, as the following examples (adapted from Brown (2008)) show. (The scales on the two axes are different).

(1)

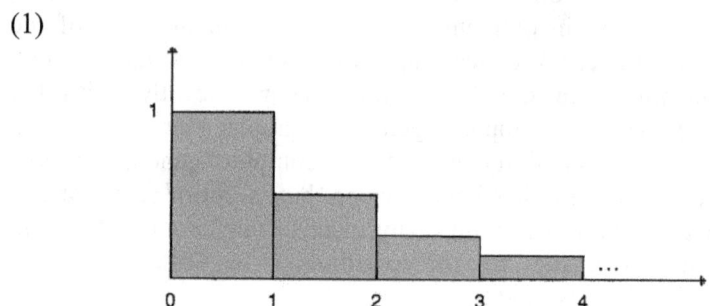

Figure 2 Graphical representation of the finite sum of the infinite geometric series with ratio 1/2.

The shaded blocks correspond to the terms of the infinite (so-called *geometric*) series

$$\sum_{n=1}^{\infty} \frac{1}{2^{n-1}} = 1 + \frac{1}{2} + \frac{1}{4} + \frac{1}{8} + \frac{1}{16} + \cdots$$

Suppose the first shaded block on the left takes 1 can of paint in order to be painted. Then the whole shaded area will take only 2 cans in order to be painted, since 2 is the sum of the series: $\sum_{n=1}^{\infty} \frac{1}{2^{n-1}} = 2$.

(2) Now consider the similar situation

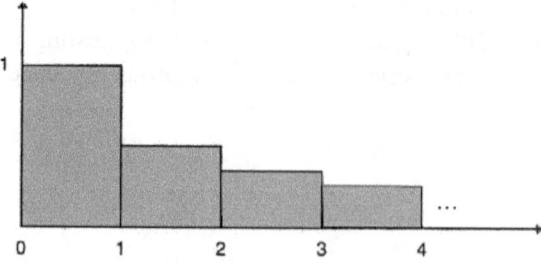

Figure 3 Graphical representation of the infinite sum of the infinite harmonic series.

corresponding to the so-called *harmonic* series:

$$\sum_{n=1}^{\infty} \frac{1}{n} = 1 + \frac{1}{2} + \frac{1}{3} + \frac{1}{4} + \cdots$$

The reader may care to guess how much paint is needed in order to cover the shaded area? A bit over 2 cans, but not more than 4, perhaps? In fact, an *infinite amount of paint is needed*, since the harmonic series diverges to infinity (that is, for any given positive number L, the finite sum becomes $1 + \frac{1}{2} + \frac{1}{3} + \frac{1}{4} + \cdots + \frac{1}{n} > L$ for sufficiently large n).

There is and there always has been a kind of visual reasoning or thinking in mathematics, that is, a sort of visual imagination or perception of figures, diagrams and symbol arrays, and mental operations on them. The question is:

- Is this visual thinking merely an heuristical or psychological aid, facilitating grasp of what is gathered by other means?

- Or does it also have epistemological functions, as a means of discovery, understanding, and even proof?

Many have argued (see bibliography) that visual thinking in mathematics is rarely just a superfluous aid; it usually has epistemological value, often as a means of discovery in different areas of mathematics, from elementary geometry to algebra and analysis. It seems that sometimes we can discern abstract general truths by means of specific images, and use visual means to help us grasp abstract structures.

Some examples of picture-proofs

Picture-proofs or proofs without words in mathematics are generally pictures or diagrams that help the reader see why a particular mathematical statement may be true, and how one could begin to go about proving it. But:

- Are picture-proofs rigorous proofs a par with more traditional verbal/symbolic counterparts, or no more than means to provide plausibility evidence or just visual clues to stimulate mathematical thought?

Let us look at some examples of picture-proofs, from arithmetic and algebra.

(3) **Theorem** For every natural number $n \geq 1$,

$$1 + 3 + 5 + \cdots + (2n - 1) = n^2$$

Picture-proof:

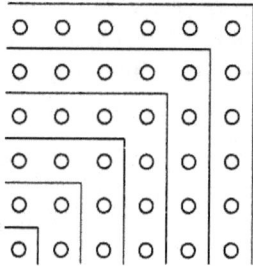

Figure 4 Graphical representation of a picture-proof of a theorem in the Theory of Numbers.

Represented in this picture is the case $n = 6$.

This is traditionally proved by mathematical induction, as is the following:

(4) **Theorem** For every natural number $n \geq 1$,

$$1 + 2 + 3 + \cdots + n = \frac{n^2}{2} + \frac{n}{2}$$

Picture-proof:

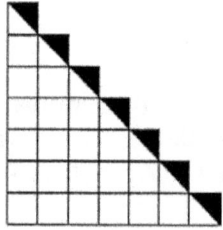

Figure 5 Graphical representation of a picture-proof of a theorem in the Theory of Numbers.

Represented in this picture is the case $n = 7$. We must now ask: Is this a picture-proof for the case $n = 7$ only, or can it in some sense be taken (as Brown (2008), p. 40 claims, arguing more extensively and invoquing Chihara (2004)) to establish the full result? Let us quote Brown on this:

> "*Consequently, I want to suggest something quite different. My bold conjecture (to use Popper's terminology) is this: Some 'pictures' are not really pictures, but rather are windows to Plato's heaven. The number theory diagram is certainly a representation for the $n = 7$ case, but it is not for all generality. For the latter, it works in a different way, more like an instrument. This, of course, is a realist view of mathematics, but not a realist view of pictures. As telescopes help the unaided eye, so some diagrams are instruments (rather than representations) which help the unaided mind's eye.*"

On the other hand, some may claim that one sees in the picture the possibility of a reiteration: for $n = 8, 9, \ldots$; the diagram can be extended to any number. More to the point, some see mathematical induction <u>encoded</u> in the diagram (Chihara), others don't (Brown).

Here is a verbal/symbolic proof by mathematical induction.

Traditional proof

Base of the induction (i.e. case $n = 1$): $1 = \frac{1^2}{2} + \frac{1}{1}$. Induction step (i.e. from n to $n + 1$, for arbitrary n): Suppose (Induction Hypothesis) that the equality holds for n, i.e. that $1 + 2 + 3 + \cdots + n = \frac{n^2}{2} + \frac{n}{2}$. We then have:

$$(1 + 2 + 3 + \cdots + n) + (n + 1) = \left(\frac{n^2}{2} + \frac{n}{2}\right) + (n + 1)$$
$$= \frac{n^2}{2} + \frac{n}{2} + \frac{2n}{2} + \frac{2}{2}$$
$$= \frac{n^2 + 2n + 1}{2} + \frac{n + 1}{2}$$
$$= \frac{(n + 1)^2}{2} + \frac{(n + 1)}{2}$$

This proves the equality of the theorem for $n + 1$. By the principle of mathematical induction, the theorem holds. Brown further comments:

> "Some mathematicians (and similarly some philosophers) are often quite adamant that induction definitely is or that it definitely is not encoded into the diagram. The common claim of some who see induction in the picture is that the picture is indeed legitimate as a proof, but it is so because of induction, not because of some Platonic reason. Others say that the picture is really a heuristic device that suggests mathematical induction, and that it is induction itself that is the genuine and legitimate proof of the theorem. These are interesting objections and I'll say three things in reply."

First, my view about pictures is two-fold: that they can play an essential role in proofs *and* that there is a Platonistic explanation for this. One version of the Kantian objection (or the it-is-really-just-induction objection) is only in opposition to the second, Platonistic, aspect. Pictures, and seeing the possibility of constructions, can still be a legitimate form of mathematical proof. Indeed, the legitimacy of pictures is upheld by the objection.

Second, the different interpretations of how the picture works are related to the distinction between potential and actual infinities. Both see the formula as holding for all $n \in \omega$.[2] But the Kantian iteration account sees ω as a potential infinity only; the Platonistic account sees ω as an actual or completed infinity. Of course, the proper understanding of infinity is an unsettled question, but classical mathematics (especially set theory) seems committed to actual infinities. So I see my Platonistic interpretation of how the picture works as being favoured for that reason over the Kantian one.

Third, we might well wonder: What has the perceived possibility of constructing a diagram of any size got to do with numbers? I certainly don't deny that we can see the possibility of indefinite iterations of the diagram, but the Kantian objection seems to assume that we know the number theory result *because* we see the possibility of iteration. I don't know of any argument for this. We could just as well claim that we see the possibility of iteration *because* we have the prior perception of the number theory result.

I shall end this brief exposition with another quote from Brown (2008), p. 47, which summarizes his thinking and project:

> *"Philosophers and mathematicians have long worried about diagrams in mathematical reasoning — and rightly so. They can indeed be highly misleading. Anyone who has studied mathematics*

---

[2][The set of natural numbers, including 0, is usually denoted ω, or $\mathbb{N}$. For $n = 0$ the equality in the theorem is trivial, since it reduces to $0 = 0$].

in the usual way has seen lots of examples that fly in the face of reasonable expectation.

(...) Picture-proofs are obviously too effective to be dismissed and they are potentially too powerful to be ignored. Making sympathetic sense of them is what is required of us."

In my opinion, Brown's book goes a long way in defending intelligently the case for visual thinking in mathematics and the use of diagrams not only as a powerful aid to intuition but also a means to establish very general theorems. He also argues convincingly that <u>in some cases</u> they are just as rigorous as any traditional verbal/symbolic proof. I am beginning to see the picture.

References

Barwise, J. and Etchemendy, J. (1991) "Visual Information and Valid Reasoning", in W. Zimmerman and S. Cunningham (Eds), *Visualization in Teaching and Learning Mathematics,* Mathematical Association of America.

Borwein, J. (2008) Implications of Experimental Mathematics for the Philosophy of Mathematics, in Gold and Simons (2008) pp. 33-59.

Brown, J.E. (2008) *Philosophy of Mathematics, A Contemporary Introduction to the World of Proofs and Pictures,* Second Edition, New York and London: Routledge.

Chihara, C.S. (2004) *A Structural Account of Mathematics,* Oxford: Oxford University Press.

Detlefsen, Michael (2008) Proof: Its Nature and Significance, in in Gold and Simons (2008), pp. 3-32.

Dieudonné, J. (1969) *Foundations of Modern Analysis,* New York: Academic Press.

Gold, Bernie and Simons, Roger A. (Eds) (2008) *Proof and other Dilemmas: Mathematics and Philosophy,* 2008, Washington: Mathematical Association of America.

—— (1964) *Algèbre linéaire et géometrie élémentaire,* Paris: Hermann.

Hahn, H. (1956) "The Crisis in Intuition", *in* J.R Newman (ed.), *The World of Mathematics,* New York: Simon and Schuster.

Hammer, E. (1995) *Logic and Visual Information,* Stanford: CSLI.

Giaquinto, M. (1994) "Epistemology of Visual Thinking in Elementary Real Analysis", *British Journal for the Philosophy of Science.*

—— *The Search for Certainty,* Oxford: Oxford University Press.

—— (2007) *Visual Thinking in Mathematics: An Epistemological Study,* Oxford: Oxford University Press.

Mancosu P., Jørgensen, K.F. and Pedersen, S.A. (Eds.) (2005) *Visualization, Explanation and Reasoning Styles in Mathematics,* Amsterdam: Springer Netherlands.

Nelsen, R. (1993-2000) *Proofs Without Words* Vols. I, II, Washington: Mathematical Association of America.

Senechal, M. (1998) "The Continuing Silence of Bourbaki — An Interview with Pierre Cartier", *Math. Intelligencer* 20(1): 22-28.

Concept Script and Diagrams:
Two Ways of Using Images in Scientific Reasoning[1]

by
Ángel Nepomuceno-Fernández[2]

Abstract

Frege can be considered with respect to modern logic on the same level as Aristotle with respect to traditional logic and Peirce is especially a philosopher that grounded his most important ideas on philosophical logic. They created different systems of symbolic logic that proved to be effective to represent scientific knowledge. In this work we compare the conceptual writing proposed by Frege and Peirce's diagrams, two forms of picturing logical structures of propositions, then graphic representations of scientific reasoning, to conclude that both points of view should be seen as complementary.

1. Introduction

Since in this work we are paying attention to some aspects of logical thought of Frege and Peirce, we dedicate this introduction to revise briefly relevant moments or figures in the history of logic that perhaps influenced such authors. To simplify, first we focus on the Scholastic Logic, which was developed in West Europe between XIII and XV centuries and achieved the most development in the Iberian Peninsula through subsequent centuries; representative cases are the Nominalistic School of Salamanca and the Coimbra School, a Jesuit school that edited the world-famous companion of Scholastic Logic called "Cursus Conimbricensis". If XIII century is considered the starting point, it should be pointed out that Schoolmen had a whole series of comments about works due to Aristotle, Porphirius and Boetius, though Saint Anselm, who had a great knowledge of the so called

---

[1] Research for this paper was supported by subventions from *Ministerio de Educación y Ciencia (Spanish Govern)*, project HUM2007/65053 and from *Consejería de Innovación, Ciencia y Empresa (Andalusian Govern)*, project P06-HUM01538. A preliminary version was presented as part of activities developed in the framework of the project *Image in science and art*, Centro de Filosofia das Ciências de Universidade de Lisboa.

[2] Dpto. Filosofía, Lógica y Filosofía de la Ciencia, Universidad de Sevilla, e-mail: nepomuce@us.es.

"logica vetus", must be underlined. Besides other previous works, the legacy of Abelard is unquestionable, he was pioneer in examining the "questions", that is to say, two perspectives that are conflicting, shown as efficient exercise of the reason for a debate and resolution of created problems. The logical legacy due to Abelard is constituted by four works where he used a methodology that was adopted by the rising universities, establishing a basis of the famous "disputations" that come to obtain what now is intended as a university degree.

During the XIII century very important figures to study medieval logic appear. In accordance with Boehner (2007), Albert classifies concepts according to predication modes (gender, species, difference, property and accident), pays attention to the last six Aristotle's categories, studies the division of the gender in species, the whole in its parts and the word in its meanings and a distinction of accidents, he tackles the problem of the proposition and its elements and the truth and falsity as properties of propositions and a rudimentary theory about contingent futures. On the other hand, he examines not only the categorical syllogism but the modal one, which is formed by modal propositions, and the mixed syllogisms, in which a premise is a proposition of fact and the other is modal, conditional and subjunctive syllogisms. He develops a proof theory that exceeds the Aristotelian point of view and is an antecedent of the "consequentiae" theory and the art of disputes.

After a basis for Scholastic Logic had been established, there were a huge amount of authors that contributed to its development. Thomas of Aquino, William of Sherwood, Peter Hispanic, Roger Bacon, etc. are outstanding known names in history of logic. All of them study subjects of logic, which is conceived in a broad sense, since they tackle as issues that are strictly logic as disciplines enclosed in the "trivium", that is to say, grammar and rhetoric. To mention only those analysis that are related with Peirce's point o view, the properties of terms came to Schoolmen to give a theory about sincategorematic words and to a distinction between "significatio" and "suppositio" and a doctrine about the consequentiae, though there is not unanimity respect to that among to all historians (Beuchot, 2002 is an important reference about this matter).

On the other hand, if we want to trace a path that leads to Frege's thought, some names are unavoidable, as Raymond Lull and specially Leibniz. The former was a Spanish monk that thought that the best way of preaching his religious ideas ought to use a special writing by mean of which concepts were able to be represented. So Christian truths would be the result of right combinations of primitive circles taken as ideograms, each one of which would stand for basic true ideas; this kind of language was called *Ars*

Magna and it might suggest the Leibniz's project of defining a *characteristica Universalis*.

According to Kneale (1972), to study Leibniz's logical thought, one should pay attention to several interests that this author held all his life, namely

1. A profound respect for traditional logic. He thought that a deductive system could be settled by means of chaining of successive syllogisms in a similar way of constructing the Euclidean geometry, though he knew that is not possible to reduce to syllogistic form disjunctive and conditional arguments.

2. A notion of *Ars Combinatoria* that implies a conception of logic as a calculus. He conceived the idea of an alphabet for human thought such that everything that could be thought would be constructed by means of combinations of elements taken from this alphabet.

3. His project of defining an ideal language, a scientific language that would be useful not only to communicate thoughts but to think, that is to say, a lingua philosophica or *characteristica universalis*. To do that, it would be necessary a suitable class of symbols and appropriate resources to express formal notions as logical connectives, predication and universality and existence. In short, a "universal mathematic" is necessary.

4. The project of gathering the principles of all sciences in an encyclopedia.

5. To settle a general science to develop the scientific method. Unlike Descartes and other modern philosophers, such general science should take into account the logical forms and the laws that govern that.

These ideas constitute the nucleus of the logicist philosophy as long as a prominent role of logic is defended and it is the discipline that can provide the foundations of mathematical knowledge. Indeed, if logicism is considered a "metaphysic program of research", almost all logical ideas defended by Leibniz are suitable in order to carry such program out. Leibniz gives a great importance to chosen symbols and in fact his symbolism to integral calculus prevails over the symbolism proposed by Newton, which become an important characteristic of logicist positions. However, the algebraic tradition shares some of such ideas with the line of thought that was begun by Frege.

2. A language for logical forms

Frege's *Concept Script*, (Frege, 1879), constitutes the first systematic treatment of classical logic that gives an account of many previous attempt of systematizing logic, from the traditional logic due to Aristotle and Scholastic to the introduction of mathematical methods proposed by Leibniz or Boole. So, such Fregean system can explain the square of logical opposition, tackle a unified theory of predication, represent the indemonstrable principles studied by Stoics and express logical relations between classes. It is also an important ingredient to settle the logicist philosophy, according to which arithmetic (then, all analytic mathematics) is based on logic.

The book is divided in three parts. The first one is called "definition of the symbols", the second "representation and derivation of some judgments of pure thought" and the third "some topics from a general theory of sequences". One does not find a general theory of signs, but Frege begins distinguishing two kinds of signs

1. Those that have a conventional meaning and can refer different objects. They are letters of the ordinary language and have an indeterminate meaning, which depends on the situation or context, but once the meaning is assigned to a letter, it is the same meanwhile the situation is the same.

2. Those that have a completely determinate meaning, that is to say, those that have the same meaning whatever the situation may be. They constitute the set of logical symbols par excellence

The main logical signs studied are the "content stroke", "judgment stroke", "conditional", "negation", "identity of content", and "generality". To represent what Frege calls 'judgments of pure thought' he devotes two paragraphs to the new notions of "function" and "argument" and how representing them. So Frege prepared the way to use formal languages in logical studies, though he had not awareness of having defined a formal logical language in the sense that it is intended nowadays. In fact Frege understood his language as a symbolization of the natural language itself, giving a new syntax in order to purify it but retaining the semantics of natural language. The starting point is the ordinary language, the vehicle for expressing propositions in general, though it has an essential ambiguity that should be avoided, then a new linguistic analysis is proposed, very different from the traditional one, which is an inheritance from the Aristotelian and Scholastic logics. Whereas the structure of an elemental proposition was explained in terms of subject and predicate by traditional logicians, Frege, who had assumed the concept of function as a fundamental notion in mathematics, conceived an elemental proposition as the result of completion of an unsaturated expression, namely the function, by means of an argument. That is to say, the proposition "Mary studies philosophy", for example, is obtained from the function "...studies philosophy" that is saturated or completed by the argument "Mary". To tell the truth, such notions are fully studied in posterior works, however they appears yet in *Concept Script* as two irreducible elements that together constitute a simple proposition, at least at a syntactical level as an incomplete part that achieves the meaning of a proposition only by addition of a proper name (an argument). So, this new analysis is in terms of function and argument that are constituent elements of any proposition, instead of predicate and subject respectively. Predicate and subject have continued being used, not in the traditional sense, but in the Fregean sense, as saturated and unsaturated expressions respectively. On the other hand, "proper noun" is understood by Frege as a linguistic expression that denotes an object, not a function, because of which now this linguistic category is made up of proper nouns (Mary, John, etc.) and definite descriptions (The Queen of United Kingdom, the author of Ivanhoe, etc.). Later Frege conceived the doctrine of "sense and reference" (Frege, 1980) in keeping with this explanation: every sign, or linguistic expression, should have a sense and a reference. The sense of any term is known by users of the corresponding language (ordinary or ideographic) and it is the vehicle to achieve the reference. What is the reference of arguments and functions and how the composition of references takes place? The reference of any noun is an object, so if the argument is a noun this is the case, but if a functional expression is in argument position, then such function has been made noun and its reference is an abstract object. The reference of every function is a concept, unless the function would be in argument position, as indicated. The reference of any judgment is "the true" or "the false" (one of the truth value, though for Frege they are abstract objects). Every sign expresses a sense and have a reference and to several senses can correspond only a reference (the reciprocal may be not. "Miguel de Cervantes" and "The author of Don Quijote" are two proper nouns, two different senses, but both of them have the same reference, namely a famous modern Spanish writer who died when Shakespeare does).

The logical form of any proposition should be found by examining its grammatical form. However, a proposition could be expressed by means of different grammatical forms, since there will be several senses. Compare that two expressions "Mary painted this picture" and "this picture was painted by Mary", both of them represent the same proposition, namely those that result of saturating the function that can be paraphrased like "...painted this picture" or "this picture was painted by..." with the argument "Mary", but both of them refer an only fact: the fact that "this picture" was painted by "Mary", independently that the corresponding proposition had been expressed in active or passive voice. In accordance with these Fregean notions, the statement "Mary painted this picture" could be seen as composed by other elements, the function "Mary painted..." and the argument "this picture". Of course this is also correct, all depend on the adopted perspective when the final product is analyzed, that is to say, two different functions and arguments may be combined in a way that the same proposition is obtained, but the relevant is that in both cases the logical form is the result of saturating a function with one argument. Another doubt could still be arisen in this expression. Are not "Mary" and "this picture" two arguments that could be replaced by "Picasso" and "the Guernika"? Are logical forms of "Mary painted this picture" and "Picasso painted the Guernika" very different farther on the meanings of such proper nouns? Another analysis may be the following: given the function that can be expressed as "...painted...", we need two arguments to saturate it and depending on the chosen pair to do that one of the following propositions can be expressed "Mary painted this picture" or "Picasso painted the Guernika".

To deal with logical forms is possible without separating "predicates" –functions that refer to concepts, so that monadic functions that only need an argument to be saturated– from "relations" –functions that refer relations, so that functions that need more than one argument to be saturated–. This is an important contribution of Frege to construct the system for classical logic. Indeed to represent schematically logical forms of propositions the rule could be to write first the functional expression followed by brackets, which represent a gap to be fulfilled when saturated, and secondly the corresponding argument(s) between such brackets. Let us see a form of representing logical forms of the last two examples

Painted (Picasso, the_Guernika)

Painted (Mary, this_picture)

But a mathematical function can be represented as algebraically as geometrically, and then certain questions arise: Could the logical structure of a proposition be represented by means of a figure of geometry? Can a set of propositions, or the set of its logical forms that are expressed as sentences of a language, have algebraic properties? Before answering that, let us see how Frege complete his language of formulas.

Frege considers that all judgments can be represented by means of an only sign, the known "⊢", which is composed by a vertical stroke and a horizontal stroke and it has to stand to the left of the signs corresponding to the content of the judgment. This is a geometric figure after all. If only the horizontal stroke is written, for example "–A", then we have a combination of ideas, represented by A, rather than a judgment that could be true or false. In other words, "–A" expresses "the circumstance that A", though there are some requirements about what can be written to the right of ⊢. In one of the example mentioned above, as active (Mary painted this picture) as passive voice (this picture was painted by Mary) express the same proposition, since both expressions share the same conceptual content. This is a first condition: signs that are on the right of a horizontal stroke must represent a conceptual content, precisely such stroke is called "content stroke", but when the vertical stroke is added, that is to say, when the "judgment stroke" is written, for example ⊢A, then A must represent a conceptual content that itself can become a judgment. Some conceptual contents cannot became a judgment like conceptual contents of common nouns (horse, house, etc.), so that in ⊢A it is necessary that A represents a proposition, since the conceptual content of any proposition can became a judgment. But if A is a proposition, it become the following judgment: "A is a fact", then the Fregean ideography will contain ⊢ as its only predicate (in its grammatical sense).

Taking into account the former notation, the exemplified propositions cause two judgments that can be represented ideographically as follows

⊢Painted (Picasso, the_Guernika)

⊢Painted (Mary, this_picture)

or, if such terms are abbreviated in a way that P represents the (dyadic) function, a is a name for Picasso, b for Mary, c for the Guernika and e for the picture, as the formulas

⊢$P(a, c)$ and ⊢$P(b, e)$,

respectively. Any simple proposition can be represented as an atomic formula of the *Concept Script* and any compound one can also be represented in accordance with compositional principles that are provided for. Previously, Frege had discovered the material implication. Specifically, he proposes that having two conceptual contents *A* and *B*, which can become judgments, there are exactly four possibilities

1. *A* is affirmed and *B* is affirmed

2. *A* is affirmed and *B* is denied

3. *A* is denied and *B* is affirmed

4. *A* is denied and *B* is denied

Then the way of understanding "if *A*, then *B*" is that such conditional is a judgment according to which the second possibility does not take place, but one of the other three does. Accordingly, its negation means just that *A* is affirmed and *B* is denied "is a fact". The concrete representation of conditional in *Concept Script* consists on a content stroke to the left of expression of antecedent that links with the content stroke of consequent (drawn larger) that is closed by judgment stroke. But Frege writes on a peculiar way: from down to up. He wanted so to show as clearly as possible the logical connections but the real effect is a very different writing, since the most common is to write the ordinary language horizontally (from left to right or from right to left) but any concatenation of conditionals, when it is a bit complex, will become visually more explicit. This bi-dimensional writing of logical forms of propositions suggests that the differences between Frege and Peirce are less than it could be seem at first sight. Is the ideographic writing a diagrammatic representation of logical connections in the sense of Peirce? Indeed followers of Frege, though heirs to his main ideas, adopted Peano's symbolic language. To simplify this presentation, we shall use of standard symbolism to classical logic (at propositional or predicate level).

Given a formula, Frege proposes a first way of composition: the negation. If we have two propositions, other clauses of composition are disjunction, conjunction and forms of combining that with negation –"and not", "neither, nor"–. Of course these must be represented in a way that the corresponding symbols affect conceptual contents, or to be more exact, conceptual contents that can become a judgment. So for unanalyzed propositions *A* and *B* and monadic functions $P(.)$ and $Q(.)$, conceptual contents that can became judgments, a Fregean symbolic representation of judgments achieved by composition of such contents may be

1. Conditional: $\vdash A \rightarrow B$

2. Negation: $\vdash \neg A$

3. Conjunction: $\vdash \neg(A \rightarrow \neg B)$

4. Disjunction: $\vdash \neg A \rightarrow B$

5. Universal quantification: $\vdash \forall x Px$

6. Existential quantification: $\vdash \neg \forall x \neg Px$

Only conditional, negation and universal quantification have their own signs, whereas the other logical constants are represented by means of combinations of that. A last sign is the identity of content, which would be expressed as $\vdash a \leftrightarrow b$. Frege was a defender of maximum rigor in logic and mathematics, however this formula is ambiguous since symbolically a and b can stand to left and right of $\leftrightarrow$ as nouns or as propositions. Whatever the case may be, $a \leftrightarrow b$ can become a judgment: if such nouns are "Camoes" and "The author of *Os lusiadas*", then "(Camoes and The author of *Os lusiadas* are the same) is a fact" is a true judgment, but if a and b are two equivalent propositions, for example, "Portugal belongs to European Union" and "Portugal is a member of the European Union", then "(Portugal belongs to European Union is equivalent to Portugal is a member of European Union) is a fact" is also a true judgment. On the other hand, though developed in other works, according to his "contextual principle of meaning", the context in which such signs occur can clarify its exact meaning. Then the mentioned ambiguity is easily resolved in contexts.

Since relations between concepts can be represented by appealing to objects that fall (or not) under concepts, the Aristotelian square of logical opposition is presented at the end of the first part as follows. To say that 'all S are P' is equivalent to say that given any object, if it falls under S, then the object falls under P, that is to say, the concept S is subsumed into the concept P; this is the sense of universal affirmative proposition (A) and its logical structure is represented by the formula $\vdash \forall x (Sx \rightarrow Px)$. The rest of propositions to give that square are

E: $\vdash \neg \exists x\, (Sx \wedge Px)$

I: $\vdash \exists x\, (Sx \wedge Px)$

O: $\vdash \neg \forall x\, (Sx \rightarrow Px)$ or $\exists x(Sx \wedge \neg Px)$

The second part of *Concept Script* is devoted to a logical calculus. As we said above, the way of writing this symbolism is vertical, so to follow every step through the calculus may be difficult, but Fregean conceptions become paradigmatic about how to define logical calculi for classical logic. Frege called such second part "representations and derivation of some judgments of pure thought". In short, it is an axiomatic calculus that covers what nowadays is known as a classical logic system. Quantification can be applied to any symbols that could be considered an argument, however given a formula in which *Pa* occurs, the class of functions that have *a* as argument could be considered. Let $\vdash A \rightarrow Xa$ be a formula in which *A* represents a subformula where *X* does not occur, this formula expresses a function –a second order function– whose argument is *X*. Then $\vdash \forall X\, (A \rightarrow Xa)$ is a well formed formula that has been obtained by quantification over the argument of such function. In this case quantification is not restricted to individual variables; this kind of unrestricted quantification is applied by Frege, so we can distinguish between the whole ideographic calculus and the first order fragment of that. Well, the first order fragment of this logical calculus is sound and complete, that is to say, every valid formula of the (first order fragment) of such calculus –judgment of pure thought in Fregean terms– is provable (by means of the calculus) and, reciprocally, every provable formula is valid.

Lastly the third part of *Concept Script* deals with some topics from a general theory of sequences. As it is known, Frege inaugurated the logicist philosophy of logic and mathematics, particularly after his book *The foundations of Arithmetic* (Frege, 1884), however the basic ideas are already sketched. Indeed the priority of logic over (analytic) mathematics is practiced since the mathematical induction principle –Bernoulli's induction, a version of such principle, rests on the formula number 81–, for example, it is a theorem of the calculus. If it is one of the main rules of inference in mathematical practices, according to its theorematic character, induction is only one among other ones that are logically founded rules as a last resort. On the other hand, the notion of "succession in a sequence" is also developed in this part, but successor is one of the most basic concepts in arithmetic and it is logically defined, in spite of discussing about whether the concept of sequence presupposes the concept of number. Later Frege modified his system and defined a new ideography in *Grundgesetze der Arithmetik* to conclude his philosophy of mathematics, which has been broadly studied by Dummet (1995), such new ideography, in any case, keeps the most essential of *Concept Script*.

To finish this section, let us summarize the most outstanding. Frege's *Concept Script* contains leading elements of a modern theory of proof, so that any scientific reasoning could be represented ideographically, that is to say, any scientific reasoning could be bi-dimensionally represented as a set of simple logical structures (one for each atomic proposition) that are linked by means of lines that indicate forms of logical connections. So certain graphics could represent logical forms of propositions, which had been extracted from the ordinary language, though graphics constitute an alternative language in the last resort. The subsequent logical languages, lineally written, gave the bi-dimensionality up but such logical connections are expressed geometrically (let us say "one-dimensionally") after all: from $\vdash A \rightarrow (B \rightarrow C)$, $\vdash A$ and $\vdash B$, for example, where A, B and C are propositions (conceptual contents that can become judgments), $\vdash C$ is obtained. Here logical connections are (one-dimensionally) represented compositionally by means of graphic elements, $\rightarrow$ to link two propositions, left-antecedent and right-consequent, and brackets to mark compound propositions, finally $\vdash$ to indicate the corresponding judgment. The rule in this case is the known *modus ponens*, in ideographic terms: from $\vdash A \rightarrow B$ and $\vdash A$, $\vdash B$ is its immediate conclusion.

3. Diagrammatic reasoning

In this section we are presenting in a schematic form the main ideas developed by the American philosopher C. S. Peirce that could serve to compare these two outstanding contributions to history of logic. In this paper it is not carried a detailed analysis of Peirce's and Frege's ideas out, it should be noted yet, since a deep textual exegesis requires an extension that is more suitable for another kind of work. Our intention is more modest: to raise more attention to both points of view, which can be seen as complementary and that could explain the extraordinary and fascinating history of contemporary logic and its importance to develop methodology and philosophy of science.

Peirce, with other scientists and philosophers (W. James, Ch. Wright, F. E. Abbot, etc.), founded the Metaphysical Club that lived from 1872 to 1875, year in which Wright died and Peirce travelled to Europe the second time. As a consequence of discussions in such circle, Peirce published a set of articles that are grouped under the generic epigraph *Illustrations on the Logic of Science*, where his main ideas about logic are synthesized. To begin, we must say that logic in Peirce is not conceivable as a discipline totally separated from knowledge theory. Precisely pragmatism considers there is no concept or proposition that could have meaning without referring to some context of action, that is to say, the knowledge has a dynamic character, which connects with the modern position in logic known as "dynamic turn" (specially visible after productive applications of logic in computer sciences). There are elements that inform all Peirce's thoughts as category theory, for example, about which it is enough to point out its three-fold division: "firstness", "secondness" and "thirdness". Our author conceives logic as semiotic, since all thoughts are produced by means of signs and logic must study general laws of signs. More specifically, logic, one of the three normative sciences, is divided in three branches, namely "speculative grammar", a general theory about the nature of signs and analysis of reasoning in its last elements; "critic", which classifies arguments and determines its validity, this is why sometimes it is called "formal logic", and "methodeutic" or "speculative rhetoric" whose main object is to study methodology of science, it is a kind of heuristic, giving the famous three stages that make up the scientific method: deduction, induction and abduction, which is his most important contribution to epistemology. In fact, modern logic, philosophy of science, computer sciences and artificial intelligence, among other disciplines, pay the greatest attention to Peircean formulations and studies about abduction. Such three branches somehow correspond with the modern linguistic disciplines that are called "syntax", "semantics" and "pragmatics", respectively. We should also point out that for Peirce mathematics is the science of discovery that investigates the realm of abstracts forms, the realm of ideal objects (Houser 1991, I), it is not based in logic, in short.

The Peircean model of semiotic analysis leads us to analyze the "semiosis" as a process in which something can operate as sign and the presence of three elements is required. Such elements are

1. Sign. This is something what represents something to someone in some aspect or capacity and it is directed to someone, creates a different sign, or perhaps a more developed one, in her mind.

2. Object. This is the extrinsic reality that is referred by the sign (somehow, it helps to determine the sign in order to its representation).

3. Interpreter. This is a new sign that is generated through the process. It is a mediator representation or interpreter of the (initial) sign.

These three notions correspond with the known categories, sign with firstness, object with secondness and interpreter with thirdness, but each semiotic category is defined taking only into account its logical position and it could change during the semiotic process. Though there could be degenerate semiotic relations, where one can dispense with any element, the three elements must be constituent of the process. Whatever the case may be, two semantic relations are essential as far as a sign is: the function of denoting or referring an object, and the function of originating an interpreter, a meaning after all. Such doctrine is based on two essential assumptions, first a suitability relation between the actual world (*ordo rerum*) and the world of ideas (*ordo idearum*) that rests on an ontological thesis, according to which the nature is governed by laws, and an epistemic thesis, according to which the nature is intelligible, both of them agree some points of Scholastic doctrines. Besides such adaptation (*ordo rerum* and *ordo idearum*), which points out a certain inheritance from the Scholastic, we should consider that Peirce had an enormous trust in scientific method, which assures a "happy end" for an unbounded semiotic process as a process in which triadic relations arise recursively. Such process causes a rational activity that may be characterized by a capacity for reflecting consciously, permitting inferences and concluding by giving a judgment.

To understand the role of logic, we have to take into account that Peirce considers that signs have a particular way of being connected to each other. Actually the logic is a "universal semiotic" that have to investigate such way of connection, so that any logical system must show correctly a semiotic process, that is to say, it should be a semiotic system (a conceptual net integrated by signs, in its mentioned triadic relation: sign-representation, object and interpreter). But a sign has three aspects, in accordance with the way of characterizing how the mark determines the corresponding referent (Peirce, 1991, II), namely,

1. It is "icon" in accordance with its own quality and it is a sign of anything that shares such quality. Since in this case, as an icon, the sign represents something because it resembles it, there must be some similarity (of image or structure) between sign and referent. A brush stroke, for example, can be a sign of the horizon or a geometric line (firstness)

2. A sign is an "index" when it is indeed connected to its object. All natural signs are of this nature: a flash of lightning is a sign of storm (secondness)

3. A sign is a "symbol" (Peirce gave name to that by means of the word "token"), not because it has a quality that is the same than a quality of an object, but it is interpreted as a sign of another sign, it can represent a generality and make that present conventionally (thirdness).

We could wonder if this is a pictorial conception of logic similar to the first Wittgenstein's conception of language. About the mentioned iconic aspect of signs, Peirce himself says

> *"I call a sign which stands for something merely because it resembles it, an icon. Icons are so completely substituted for their objects as hardly to be distinguished from them. Such are the diagrams of geometry. A diagram, indeed, so far as it has a general signification, is not a pure icon; but in the middle part of our reasonings we forget that abstractness in great measure, and the diagram is for us the very thing. So in contemplating a painting, there is a moment when we lose the consciousness that it is not the thing, the distinction of the real and the copy disappears, and it is for the moment a pure dream – not any particular existence, and yet no general. At that moment we are contemplating an icon"* (Peirce, 1991, I: 226)

An icon makes reference to an object by virtue of its similarity and it has visual qualities, which does not imply to lose its indexical and symbolic qualities. Iconic signs, in accordance with the way of firstness in which they participate, could be diagrams. A diagram is not merely a pure icon but, as long as iconic qualities of reasoning were expressed in its form, all valid inferences are diagrammatic. In general, a diagram is a representation that is not a merely linguistic one, but an external representation of certain relations (Dau, 2006). In fact, diagrams are icons by means of which connections between rationally related objects can be represented, so to represent any situation by means of a diagram, there must be a correspondence between both of them in a way that the own characteristic of such diagram just reflects the main properties of the represented situation.

On the other hand, the reasoning par excellence is the mathematical one, which basically consists of constructing a diagram according to a general process, observing certain relations between parts of such diagram, showing that such relations hold for diagrams of the same kind and affirming this conclusion in general terms. Definitely, logic has to be graphical or diagrammatic and "existential graphs" are one form of diagrammatic reasoning, perhaps the most important, which may play a central role in the Peircean philosophy of logic and is proving to be interesting because of its modern applications in certain field of computer sciences (Peirce 1991, II: pp. 206 and 279). The thing is that inferential aspects of mental activities and the corresponding representations could be studied by certain diagrams. In short, there are three types of existential graphs (Pietarinen 2006; Zeman, 2002), which correspond to parts of logic as follows:

1. Alpha graphs: propositional logic

2. Beta graphs: fragments of predicates logic with identity

3. Gamma graphs: they includes
   a. modalities (in general: logical, epistemic, etc.)
   b. higher-order reasoning
   c. meta-logical graphs (graphs of graphs, in a way that the corresponding syntax may be encoded)
   d. non-declarative assertions (commands, interrogatives, emotions, etc.)

Alpha graphs permit a representation of classical propositional logic. Any blank sheet, the "sheet of assertion" (or place where a graph is constructing), begins as an empty graph. Letters represent propositional variables, then two letters P and Q written in a sheet expresses its conjunction $P \wedge Q$. A circle that contains P represents the negation of such propositional variable. Now, by combination of such representations, the rest of propositional connectives can be represented and certain rules of inference can be defined, so that propositional logic may be studied in these terms. On the other hand, by means of beta graphs, as it is said, important fragments of predicate logic with identity become represented. As an example, let us see the traditional logical opposition square. First, an object is represented into the sheet by means of a dot (small letters can be used to name such objects), which affirms that the object exists; two dots put together express that both are identical, but the indication of being different, as negation of identity, is expressed by means of a circle that contains such dots. The simple (Boolean) square of opposition is obtained as follows. A capital letter P written into a sheet of assertion and a segment that touches P

(like a small tail) expresses the categorical proposition "something is P"; if P is into a circle that is cut by the line, then it is intended as "something is not P"; when the former is included into a circle we obtain a representation for "nothing is P", which is equivalent to "it is not the case that something is P", but if a new circle contains the representation of "something is not P", then we express the universal categorical proposition "everything is P". On the other hand, two capital letters P and Q written in a sheet of assertion and a line that units such letters is the diagrammatic expression of "some P is Q", that is to say, "there is x such that x is P and Q" and, by tracing a circle to contain that, we express its negation: "no P is Q", then Aristotelian square of opposition could be completed.

These logical operations are one form of diagrammatic reasoning, though a diagram can be as geometric as algebraic, then its characteristic are algebraic after all. Formulas of an algebraic representation of logical structures, as long as from its manipulation new properties may be found, can be seen as diagrams, so that any (symbolic) deduction implies the construction of a diagram

> "...the relation of whose parts shall present a complete analogy with those of the parts of the object of reasoning, of experimenting upon this image in the imagination, and of observing the result so as to discover unnoticed and hidden relations among the parts. For instance, take the syllogistic formula "all M is P; S is M ∴ S is P". This is really a diagram of the relations of S, M and P" (Peirce 1991, I: 227-228)

Indeed syllogism (and dialogism) is tackled from a more algebraic point of view. Peirce considers that the general type of inference is that one which expresses sequentially premises, a sign of illation (in this case, ∴) and conclusion. Then "$P \therefore C$" should be interpreted as expressing certain reasoning whose premises are of the class P and its conclusion is C. But adopting a notation similar to one due to De Morgan, concretely $\prec$ and $\preccurlyeq$, $P_i \prec C_i$ represents the truth of its leading principle, a specific reasoning of such class, P_i is any one of the class of premises and C_i the corresponding conclusion, but $P_i \preccurlyeq C_i$ says that such reasoning is not valid, so that, in general, $\prec$ stands at first for a logical implication. Peirce writes about it that the form $P_i \preccurlyeq C_i$ implies either

1. that it is impossible that a premise of the class P_i could be true,

2. that every state of things in which P_i is true is a state of things in which the corresponding C_i is true.

However, that can be understood in a way that a form of 'deduction theorem' is implicit. In general, given a deductive system Σ, a set of sentences P and sentences A and B, this (meta)theorem says that if B is Σ-deduced from P with A, then "if A, then B" is Σ-deduced from P only. On the other hand, why not to use $\therefore$ and $\prec$ with the same meaning? The first one in $P \therefore C$ is a mere representation of an inference, its leading principle is not expressed, but $P_i \prec C_i$ contains all that is necessary besides premises to justify the conclusion. Besides that, $\prec$ can also be used to express relation whose terms are subject and predicate, or antecedent and consequent. Then, by using marks to indicate forms of negation (or complementation), the categorical proposition (forms A, E, I and O) can be represented as follows

A: $a \prec b$ Every a is b

E: $a \prec \bar{b}$ No a is b

I: $\breve{a} \prec b$ Some a is b

O: $\breve{a} \prec \bar{b}$ Some a is not b,

which is the result of modifying lightly the De Morgan's notation

A: $A \prec B$ Every A is B

E: $A \prec \bar{B}$ No A is B

I: $A \precsim \bar{B}$ Some A is B

O: $A \precsim B$ Some A is not B.

In the Peircean notation, it should be noted, the form $A \precsim B$ is changed by $\breve{a} \prec \bar{b}$, that is to say, $\breve{A}$ expresses "some-A", and $\bar{B}$ represents "not-B". It is not a simple substitution of signs, but it is also a change in its semantic perspective, as it is said by our author,

> *"There is, however, a difference between the senses in which these propositions are here taken and those which are traditional; namely, it is usually understood that affirmative propositions imply the existence of their subjects, whiles negative ones do not. Accordingly it is said that there is an immediate inference from A to I and from E to O. But in the sense assumed in this paper, universal propositions do not, while particular propositions do, imply the existence of their subjects"* (Peirce, 1991, I, p. 208)

The study of logic of relatives was made by Peirce within the framework of Boolean logic, which makes up of contributions due to important logicians as De Morgan, Schröder, etc. (Kneale, 1972). In this theory, Peirce introduces two new symbols, namely Π and Σ, interpreted as "for all" and "there exists" respectively. So if a is a relative term, by means of indices i, j the proposition "i is the relate an j the correlate of the relation a" (or "i and j are a-related") is represented by the symbolic expressiona a_{ij}, meanwhile $\Pi_i a_{ij}$ and $\Sigma_i a_{ij}$ say that "all i's are a-related with j" and "there exists i such that it is a-related with j" respectively. If we rummaged more in Peirce's writing, we would find that his quantification theory could be extended to be equivalent to that one ideographically proposed. In some ways, diagrammatic representations are elements of a language different from the (doubly articulated) natural language, but a language after all.

By summarizing, logic was conceived by Peirce more in the sense of a general theory of scientific knowledge than today's classical logic. As universal semiotic, logic must deal with semiotic processes taking into account the aspects of signs. So topics studied by Peirce, as a qualified expert on Scholastic thought, cover somehow what today is known as philosophical logic. On the other hand, since the mathematical reasoning is paradigmatic, the most basic and it is diagrammatic, logic itself should be graphic and pay attention to existential graphs, alpha, beta and gamma graphs, which are the most important forms of diagrammatic reasoning. By means of alpha and beta graphs many reasonings and logical connections can be represented, particularly the square of opposition (Boolean and Aristotelian) and propositions that contain quantifications operators. A graphical logic is an algebraic logic (and vice versa).

4. Concluding remarks

If we want to compare the contemporary Frege and Peirce, a first *question* arises: Were Peirce and Frege aware of the other's conceptions?

Taking into account what Frege and Peirce wrote explicitly, a first answer to such question must be negative. Of course, a radical disagree turns up: their different points of view about the relationship between logic and mathematics; if Frege is the founder of logicism, Peirce rejects any reducibility of mathematics to logic. If you distinguish two logicist thesis, one according to which logic has epistemological priority over mathematics and another that defends that the priority has an ontological character, even then Peirce would turn out contrary to logicism. About that Haack (1993) proposes a similar distinction of logicist thesis: (L1) Formal thesis. All the special concepts of mathematics are definable in pure logical terms, and all theorems of mathematics are derivable from purely logical principles; (L2) epistemological thesis. The epistemological foundations of mathematics lie in logic. She argues that Peirce would accept L1 and would reject L2, but Houser (1993) denies that on the basis of two arguments, first that L1 cannot be accepted without L2, second that there is no textual evidence about Peirce's sympathy with L1.

Frege and Schröder maintained a strong controversy as a result of the review of *Concept Script* by the second, where he seemed rather critic with respect to aspects of the Fregean system. Since Schröder and Peirce were in harmony, so to speak, a comparison between Schröder and Frege could be seen as a way of comparing Frege and Peirce (Hawkins, 1993, is devoted to study that). But our two authors belong to two different traditions in the development of modern logic: Peirce worked in the line of the algebra of logic but Frege created the first known system of classical logic (in the sense of mathematical logic tradition); the former had studied logics of Aristotle, Scholastic, Boole and De Morgan meanwhile Frege's conceptions would be related with Lull's ideas and Leibniz's project (though Leibnizian ideas also influenced the algebraic tradition). Whatever the case may be, such traditions are not incompatible but complementary.

Both, Frege and Peirce, stressed the necessity of representation of logical forms by means of non linguistic recourses (different from those of the natural language), by using some kinds of "image", algebraic models or diagrams. *Concept Script* provides a bi-dimensional system of representing logical forms, later reduced to a lineal writing with abstract symbols for logical constants, in which any argument can be symbolized, taking a semantics that is not contradictory with an interpretation of the system in terms of modern model theory, that is to say, with semantic notions in the sense of Tarski. From a Peircean point of view, easily could be accepted such representations as diagrams and, of course, the system taken as an algebra. On the other hand, Peirce accepted the distinction "subject/predicate" but he does not remain rooted in the traditional logic, on the contrary we can analyze a proposition like "*S* is *P*" in terms of his logic of relatives so that it can be seen as inclusion, not only as a simple copula between a subject and a predicate. Both authors achieved independently a theory of quantification. In the last resort, it seems that there is no mutual influence between them, but their respective forms of representing inferential contexts can be seen as a way of modeling such contexts, which may be the better realization of the Leibnizian idea of a characteristica universalis (Peckhaus, 2004) and how should be used images to picturing scientific inferences.

5. References

Beuchot, M. (2002): *Estudios sobre Peirce y la Escolástica*. Cuadernos de Anuario Filosófico Universidad de Navarra, núm. 150. Pamplona.

Boehner, Ph. (2007): *Lógica Medieval. Un bosquejo de su desarrollo de 1250 a 1400*. Trad. F. Álvarez Ortega. Universidad Iberoamericana, México.

Dau, F. (2006): "The Role of Existential Graphs in Peirce's Philosophy", in Øhrstrøm, P.; Schärfe, H.; Hitzler, P. (Eds.) *Conceptual Structures: Inspiration and Application: Contributions to ICCS 2006*, Aalborg University Press.

Frege, G. (1879): *Begriffsschrift, eine der arithmetischen nachgebildete Formelsprache des reinen Denkens*. Translation: *Concept Script, a formal language of pure thought modelled upon that of arithmetic* by S. Bauer-Mengelberg in J. Van Heijenoort (2002) *From Frege to Gödel: A Source Book in Mathematical Logic, 1879-1931*. Harvard University Press.

Frege, G. (1884): *Die Grundlagen der Arithmetik: eine logisch-mathematische Untersuchung über den Begriff der Zahl*. Translation: J. L. Austin (1980) *The Foundations of Arithmetic: A logico-mathematical enquiry into the concept of number*. Northwestern University Press.

Frege, G. (1980): P. Geach and M. Black (Eds. and trans.) *Translations from the Philosophical Writings of Gottlob Frege*. Rowman & Littlefield Pub Inc.

Dummett, M. (1995): *Frege: Philosophy of Mathematics*. Harvard University Press.

Haack, S. (1993): "Peirce and Logicism: Notes Towards an Exposition'", *Transactions of the Charles S. Peirce Society*, vol. XXIX, No. 1, 33--57.

Hawkins, B. S. (1993): "Peirce and Frege, A Question Unanswered", *Modern Logic*, Volume 3, Number 4, 376-383.

Houser, N. (1991): "Introduction I, II" in Peirce (1991), xix-xli and xvii-xxxviii.

Houser, N. (1993): "On "Peirce and Logicism". A Response to Haack", *Transactions of the Charles S. Peirce Society*, vol. XXIX, No. 1, 58--67.

Hoffmann, M. H. G. (2007): "Cognitive Conditions of Diagrammatic Reasoning", Georgia Institut of Tecnology, Working Paper Series 24 (Avalaible at http://hdl.handle.net/1853/23809 [30/06/2009])

Kneale, W. & M. (1972): *El desarrollo de la lógica*. Trans. Javier Muguerza. Editorial Tecnos, Madrid.

Peckhaus, V. (2004): "Calculus ratiocinator versus characteristica universalis? The two traditions in logic, revisited", *History and Philosophy of Logic*, vol. 25, Issue 1, 3-14.

Peirce, C. S. (1991): *The Essential Peirce, Vols. I, II*, N. Houser, C. Kloesel (Eds.). Indiana University Press, Bloomington.

Pietarinen, A. V. (2006): "Peirce's Contribution to Possible Worlds Semantics", *Studia Logica* 82, 345–369.

Zeman, J. J. (2002) *The Graphical Logic of C.S. Peirce*, dissertation, University of Chicago (Avalaible at http://web.clas.ufl.edu/users/jzeman/, [30/06/2009]).

Diagrammatic Thinking in Physics

by
J. R. Croca

Abstract

Some examples of the deep interdependence and consequently of necessity of using diagrammatic thinking in science, in particular in fundamental quantum physics, are presented.

1. Introduction

Diagrammatic thinking has been ever since widely utilized with great profit in physics. The main reason for this large utilization results from the absolute necessity of having a clear vision of the problem we want to solve. In general, the numbers of solutions of the differential equations describing the physical situation are huge not to say infinite in certain cases. Still the ones with real physical meaningful are very few. In order to get the feeling to choose the right solutions the diagrammatic thinking is most useful. On the other hand there are very taught problems that we need first to visualize in order to arrive at a possible tentative solution. On the other hand, diagrammatic thinking helps us in the selection, systematization and organization of the pertinent information and of our ideas allowing us eventually to size and possible understand the concrete real physical implications of a chosen solution.

Some examples taken mainly from quantum physics shall help to understand the importance and magnitude of this way of making science.

2. Nonlinear master equation

The fundamental equation for the nonlinear quantum physics has the form

$$-\frac{\hbar^2}{2m}\nabla^2\phi + \frac{\hbar^2}{2m}\frac{\nabla^2\phi\phi^*}{\phi\phi^*}\phi + V\phi = i\hbar\frac{\partial\phi}{\partial t} \qquad (1)$$

a solution for this equation when the potential V is constant can be

$$\phi = Ae^{-\frac{(x-vt)^2}{2\sigma_0^2} - i(kx-\omega t)} \qquad (2)$$

Looking directly at this mathematical expression one does not gather the full meaning of this solution. Still if we make the graphical representation of it the full meaning and possible implications of it come automatically out.

Fig.1 – Graphic representation of the solution of the nonlinear master equation.

This picture aims at describing roughly the theta wave a fundamental constituent of the complex quantum entity. Yet a more complete solution describing not only the surrounding theta wave but also the acron, that is the complete quantum entity in one instant of time, also a solution to the nonlinear master equation, can be represented in three dimensions.

In the following picture we can see the very high energetic indivisible acron and its surrounding theta wave.

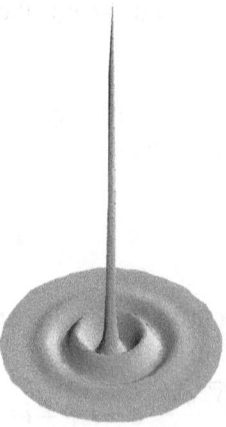

Fig.2 – Graphic representation of the complex quantum particle.

3. The unity of physics

In this case, the diagrammatic representation is of great help not only in the deep understanding of the overall situation but also in allowing us to make new inferences and connections not previously known.

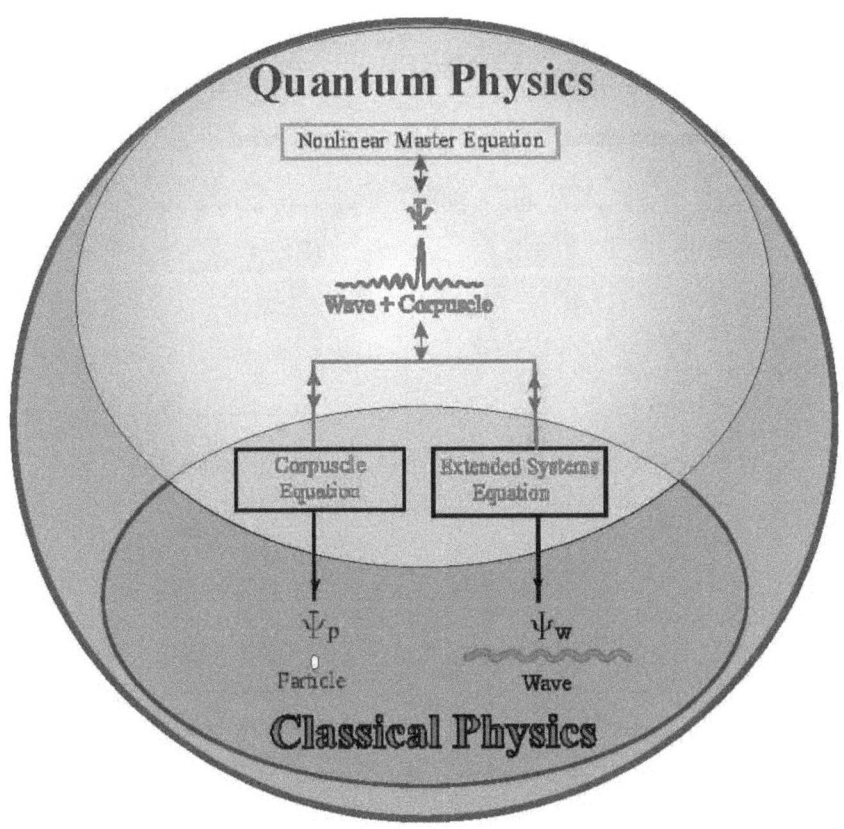

Fig.3 – Diagram for the unity of physics.

From the diagram one can easily gather that from quantum physics, from the nonlinear master equation Eq.1, by splitting it is possible to arrive at classical physics. Inversely, by fusion one arrives at quantum nonlinear physics. This diagram clearly shows the interdependence between quantum and classical physics.

4. Quantum uncertainty relations

The derivation of Heisenberg uncertainty relations not only can be better understood with the help of a diagram, but also it allows us to size its deep meaning showing that Heisenberg relations are, from the mathematical point of view, a mere consequence of Fourier ontology.

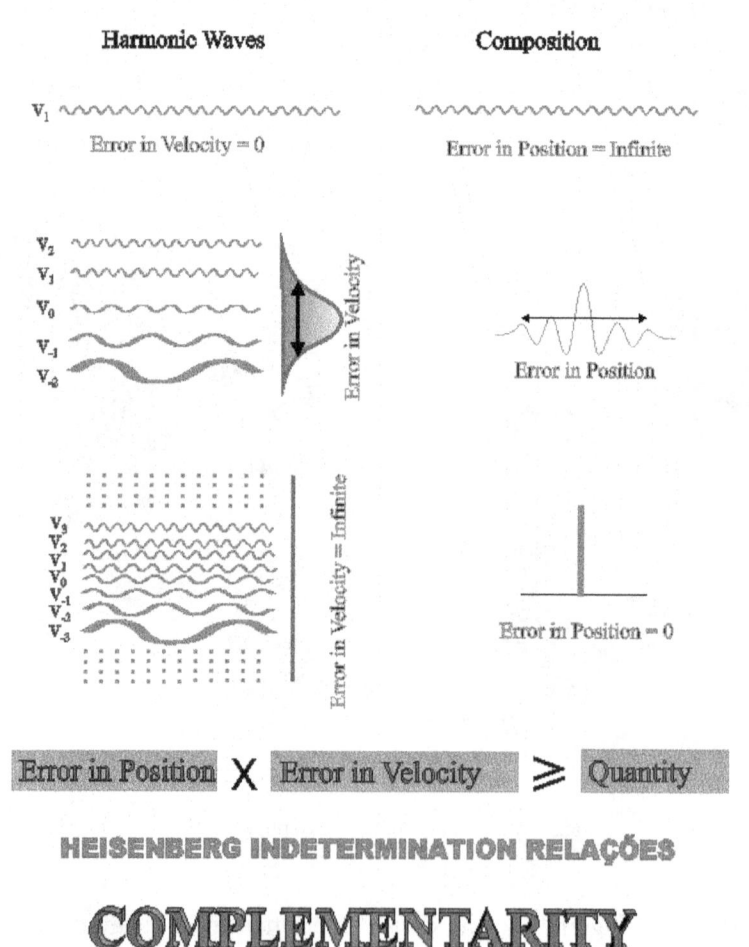

Fig.4 – Diagram for Heisenberg relations.

From the diagram we see that when the error in the velocity is zero the error in the prediction of the position is infinite. This is a direct consequence of Fourier ontology where only the infinite harmonic plane waves, the Fourier waves, do have a pure perfect frequency. On the contrary, if the error in the determination of the position is zero, meaning that we can predict with absolute precision the position of the corpuscle, then the uncertainty in the velocity is infinite.

The mathematical expression for Heisenberg relation in terms of momentum and position are

$$\Delta x \geq \frac{h}{\Delta p_x} \qquad (3)$$

that can be graphically represented

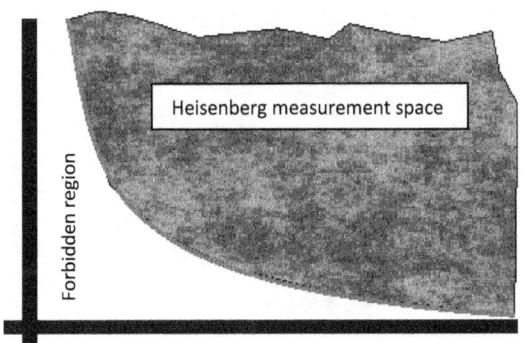

Fig.5 – Heisenberg measurement space.

To comprehend the derivation of the more general set of causal quantum uncertainty relations which have the mathematical form

$$\Delta x^2 = \frac{h^2}{\Delta p_x^2 + \dfrac{h^2}{\sigma_0^2}} \qquad (4)$$

another diagram will be of great help.

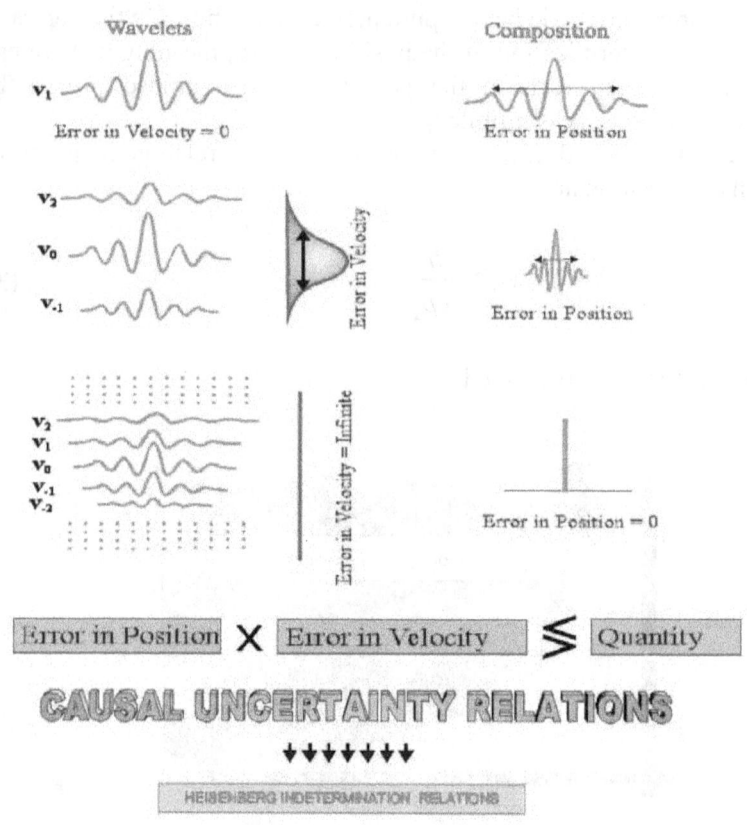

Fig.6 – Diagram for deriving the general uncertainty relations.

The general uncertainty relations space or the wavelet space can best be understood with the help of graphical representations of expression (4)

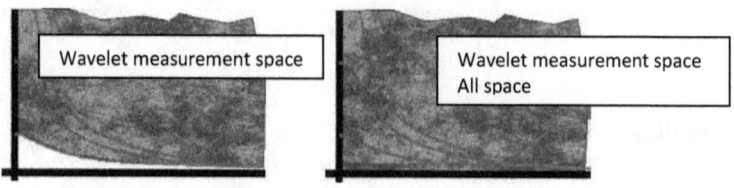

Fig.7 – Graphic representation for the general uncertainty relations.

From the diagrams not only we could infer the deep meaning of both Heisenberg relations and the general uncertainty relations but also could think on possible empirical tests for them. In fact in the next section the empirical test for the falsification of Heisenberg uncertainty relations is present also with the help of a diagram.

4.1. Falsification of Heisenberg relations

The next diagram clearly shows how the common practice of the working of the superesolution microscopes falsifies the general validity of Heisenberg relations.

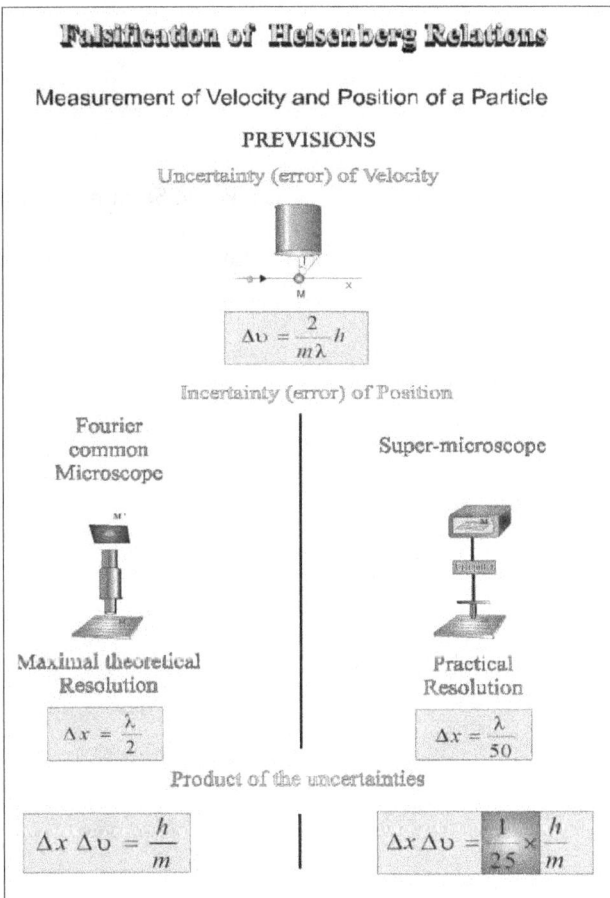

Fig.8 – Diagram for the falsification of the general validity of Heisenberg relations.

5. Conclusion

It was shown with some examples taken from nonlinear quantum physics that diagrammatic thinking is indeed a very good tool to help us to understand the deep meaning of very complex quantum phenomena and consequently to infer further useful physical relationships.

References

K. Popper, Quantum Theory and the Schism in Physics, Hutchinson, London, 1982.

W. Heisenberg, Physics and Philosophy, Harper, N.Y. 1958.

E. Schrödinger, Science, Theory and Man, Allen and Unwin, London, 1935.

N. Bohr, (1928). Como Lectures, Collected Works, Vol. 6, North-Holand, Amsterdam, 1985.

J.R. Croca, Towards a Nonlinear Quantum Physics, World Scientific, London, 2003.

J.R. Croca and R. N. Moreira, Indeterminism Versus Causalism, Grazer Philosophische Studien, 56(1999)151.

J.R. Croca and R.N. Moreira, Dialogos sobre física quântica, dos paradoxos a não linearidade, Esfera do caos Editora, Lisboa, 2008.

B.B. Hubbard, The Word According to Wavelets, A. K. Peters Wellesley, Massachusetts, 1998.

Diagrammatology, Scenographic Media, and the Display Function of Art.

by
Matthias Bauer

Abstract

Diagrams display relations and proportions. Diagrammatoidal reasoning is productive when a thought experiment is performed upon a diagram. This procedure is not restricted to logic. Rather, the display function of art is closely connected to thought experiment and the exemplification of theorematic deductions. A brief investigation of Rembrandt's famous painting The Night Watch, of Peter Greenaway's Nightwatching-project, and of Oliver Stone's feature film JFK enriches our practical understanding of scenographic media, of diagrams and of the creative imagination.

Diagrams display relations and proportions. Beside those diagrams that appear in a material medium (a sketch in a book, a floor plan, or a city map) mental diagrams function as media of thought experiment and productive thinking. What both types of diagrams have in common is the display function: the layout of relations and proportions allows not only representing a structure, but also altering the structure and getting further insights. Consequently, Peirce's concept of diagrammatoidal reasoning leads to the discrimination of two types of deduction:

> "A Corollarial Deduction is one which represents the conditions of the conclusion in a diagram and finds from the observation of this diagram, as it is, the truth of the conclusion; a Theorematic Deduction is one which, having represented the conditions of the conclusion in a diagram, performs an ingenious experiment upon the diagram, and by the observation of the diagram so modified, ascertains the truth of the conclusion." (CP 2.667)

In any case, the display function of (material or mental) diagrams demands an interplay of observation and imagination. Without this interplay, the conclusions drawn from a diagram were not reliable. Evidence, then, is the focus of shared attention when diagrams are used as tools for intersubjective reasoning and argumentation. Corollarial and theorematic deductions are convincing only when they meet the criteria of reliability –

and the test of reliability always involve inductions. But again, there are two types of inductions: empirical inductions based on experience and inductions performed according to the Pragmatic Maxim:

> *"Consider what effects, which might conceivably have practical bearings, we can conceive the object of our conception to have. Then, our conception of these effects is the whole of our conception of object."* (CP 5.402)

While empirical inductions proof the actual, observable truth of a conclusion, the conceivable bearings of a conception strongly resemble the imaginary results of a theorematic deduction. Moreover, some inductions resemble another type of inference, which is not empirical but hypothetical: the invention of a theory to explain facts which otherwise would remain inexplicable. The procedure of creative scientific thinking, then, starts with the observation that some phenomena can not be explained. To solve the problem, a theory is developed which makes this mysterious facts deducible. Of course, both the hypothesis and the deductions must be proven inductively. Since the whole procedure can be schematized, the interplay of hypothetical theory and empirical proof is a good example for diagrammatoidal reasoning. But the scope of this concept goes far beyond logic. In fact, the whole idea of sign interpretation (semiosis) is based on this concept. This becomes quiet clear in a Letter Peirce has written to Lady Welby in 1904:

> *"A sign therefore is an object which is in relation to its object on the one hand and to an interpretant on the other hand in such a way as to bring the interpretant into a relation to the object corresponding to its own relation to the object."* (Semiotics and Significs, 1977, Letter written 10-12-1904).

According to this definition, a sign causes the imagination to produce not only what Ferdinand de Saussure has named "image conceptionel" ("Vorstellungsbild"), i.e. that image of the sign's object which Peirce has called "interpretant". To fulfill this task the imagination has to perform a mimetic act since the de-sign of the image or interpretant is shaped by the way the object is re-presented by the sign. Reference, therefore, depends on perspective: the image is adaequate, only if the interpretant is brought into a relation to the object corresponding to the sign's relation to the object. What does this mean? First of all: logic is tied up with aesthetics. Second: any inference based on a certain interpretant (or chain of interpretants) follows the outline or layout of the object's de-sign. Therefore the imaginary make-up of the object as well as the conception of its meaning is diagrammatic operations. Otherwise the notion of correspondence would be senseless. If the interpretative process is structured by the object's de-sign in such a way that the chain of thought follows the given outline of relations, any departure from this outline (i.e. any transgression from reproductive to productive imagination and thinking) alters the meaning of the sign and enriches the conception of the object. Obviously, this is what a theorematic deduction does, thereby producing a need for further pragmatic considerations, namely inductions.

But, as every scholar of Peirce's semiotics knows, the whole process starts with abduction, and abductions operate on two levels: the level of perceptual judgment and the level of productive thinking. Even the notion that a certain phenomenon should be considered as a sign is an abduction, a judgment about a percept, which might be false. But if this looking at the phenomenon leads to the imagination of an object, the abduction becomes the basis for further explorations starting with corollarial deductions. If not interrupted by new phenomena, which urgently demand for a shift of attention, these corollarial deductions exhaust the object's de-sign, respectively the display function of the sign. Further insight depends on the alteration of this de-sign, i.e. on theorematic deductions. Of course, there is no unmediated transgression from reproductive to productive thinking. Rather, abduction re-enters the process where corollarial deduction ends and theorematic deduction starts. The performance of an ingenious experiment, of course, is not a deduction or induction, but an abduction. This becomes evident by two complementary definitions. The first proposes that abductions, corollarial deductions and inductions are intertwined, because: "Abduction consists in studying facts and devising a theory to explain them." (CP 5.172)

The examination of phenomena described by Peirce as "studying facts", must be a task fulfilled by either inductions or corollarial deductions, since it is a matter of observation. It is only the "devising of a theory to explain them", that fits the second definition of abduction which Peirce has proposed:

"The surprising fact, C, is observed; but if A were true, C would be a matter of course. Hence there is a reason to suspect A is true." (CP 5.189)

According to this definition, one has to distinguish the act of "devising a theory" which is an abduction, and the act of explanation which consists of deductions, pragmatic considerations, and empirical inductions. The devise of a theory based on the assumption that A might be true is the only creative act while all the other operations are controlled by observable, factual relations, i.e. by the diagrammatic logic of de-sign, lay-out and display.

This logic is not restricted to verbal communication, symbolic discourse, or narrative text. It can be found at work in paintings, feature films and other scenographic media as well. A relational make-up of phenomena, events etc. is required only. As long as there is a scheme of action, a plot or a rule of configuration to observe, there is ground (and motivation) for creative thinking involving hypothetical and empirical, theorematic and pragmatic considerations.

The aim of the following pages is to exemplify the display function of art and to show how scenographic media trigger and conduct diagrammatoidal reasoning. The first example illustrates how a soundless picture conveys a roll call, how the image incorporates a plot, and how a thought experiment is laid out in the scenographic display of shape and color, relation and proportion.

1. Example: Rembrandt's The Nigh Watch

Fig. 1 - The Night Watch.

Everybody knows this famous picture, painted by Rembrandt between 1639 and 1642. At this point of time, nobody called it The Night Watch. This title was not used before 1797. The only contemporary description was found in the family album of Frans Banning Cocq, the black, majestic figure in the centre of the canvas: "The young Captain von Pumerlandt commands his lieutenant von Vlaardingen to send out the company." The captain is Banning Cocq himself; his lieutenant was named Willem van Ruytenburgh and Vlaardingen. According to the contemporary description, a speech act rules the action which is shown in the painting. It is not only a group portrait (image) but the diagram of a certain chain of action: command – communication – configuration. After the lieutenant has conveyed the roll call, the soldiers, still in a state of confusion, have just started to build up a marching order.

Note the synaesthetic function of the diagram: The speech act which the painter is not able to represent as an acoustic sensation is inferred by the layout of three different points of time on the canvas (command – communication – configuration).

Of course, there is more to the picture than meets the eye, for example the auto-reflexive dimension of Rembrandt's painting: the configuration of the military group and the formation of the figures on the canvas are caught up in the same diagram. One might imagine, that the roll call was not uttered by the captain but by the artist in order to give the somewhat static genre of group portrait a sense of drama. "Line up in such a way that the spectator gets the impression of people reacting to a roll call." By this order, Rembrandt achieved his artistic goal: naetuerellste beweechgelhichkeit. – The painting depicts an action, a strong movement which moves the spectator.

According to this frame of reference, there is a blending of two plots in the painting, both of them deducible from the same diagrammatic design. The first one is developed by corollarial, the second by theorematic deduction.

Moreover, there is at least one more drama hidden in the layout of the painting. The saying goes that Rembrandt used a very special manual, the so called Wapenhandelinghe (Handbook of Weapons), which was published already in 1607. In this book several diagrams were used to show how guns should be handled before and after shooting. Some of them are reproduced in Rembrandt's painting: loading a musket, aiming at a target with a musket, inspecting the musket after it has been fired.

2. Example: Greenaway's Nightwatching project

Some years ago the British painter and film director Peter Greenaway started a project called Nightwatching, containing a performance to celebrate Rembrandt's 400th birthday in 2006, a catalogue, a screen play and a feature film, all of them based on the idea, that a so far undiscovered crime is hidden in the famous group portrait.

A closer look at the scenographic media used by Peter Greenaway might be useful to learn something about the creative use of diagrammatic reasoning, about the interplay of story telling and thought experiment and about the display function of art, closely related to theorematic deduction, abduction and – sometimes – induction.

Fig. 2 - The Night Watch detail.

Greenaway's abduction is very suggestive: "Look behind the heads of these two engrossed gentlemen. A gun barrel joins them. And the barrel is firing a shot. Its orange flame issuing from the barrel-end aligns itself with Willem's hat-brim, and the musket-barrel's smoke aligns itself with the white feather in Willem's hat." (Greenaway, 2006a, p. XV)

If this conjecture is at least plausible, probably there is another story to be told – the story of a group of armed men assembling to hide a shot. What does the spectator see? A military exercise, an accident or even an assassination? Suspicion arises and triggers abduction. Can we read the plot of a conspiracy into the picture? Can wee see a diagram in the image that displays a crime?

In fact, the gun shot ist not the only mystery in the painting. Working like a detective, Greenaway discovers some other riddles and wonders how they can be considered as related clues.

> "The Night Watch is not a snapshot containing incidental or coincidental information. Every shadow, gesture, contrast, detail, drama, placement and event is considered and there presented for purpose. Some have counted fifty-one mysterious or ambiguous or problematic events in the painting. But they are only mysterious or

ambiguous or problematic if they are not considered to be comprehended as a whole in association with the murder and the conspiracy." (Greenaway, 2006a, p. X)

The attempt to comprehend as a whole is crucial because otherwise details would be dispelled. Diagrammatic understanding is covered by a principle of continuity. This principle can rule the collecting of marks laid out on a canvas (in a spatial order) as it can rule sign events laid out in a time based medium like drama, feature film, or video. The assumption that there is a scheme of relation, probably connected with a plot of murder and conspiracy, transforms the image into a genre scene and triggers further observations and associations. For example:

"[...] Willem van Ruytenburgh carries the halberd in such a fashion as to infer that its bold, blunt-blades head becomes a substitute phallus with a strong suggestion of testicles. And apparently this shadow of a hand with splayed fingers, is decidedly groping towards Willem van Ruytenburgh's belly and genitals." (Greenaway, 2006a, p. XIII)

The significance of hand and shadow, belly and weapon is enforced by the doubling of the spot light-effect Rembrandt used. The yellow van Ruytenburgh is mirrored by another character on the other side of Banning Cocq – the only female character in the group portrait. The impression is unavoidable that the lieutenant, apparently a small man, is an intimate to the big male in the centre of the canvas. Obviously there is a relationship between two men, a relationship of power and sexuality.

"Another curiosity: Why does Frans Banning Cocq carry a right-handed glove in his right gloved hand, when surely it is his left hand that is gloveless?" (Greenaway, 2006a, p. VIII)

This question leads back to another picture:

"Rembrandt's involvement in all this activity goes back a few years. In 1639 he had painted a full-length portrait of Andries de Graeff, probably the richest man in town, certainly a man of power and influence. [...] But the painting was not appreciated and it was sent back to Rembrandt without payment. [...] Rembrandt sued, and won. Andries had to pay up. In the painting, there is a right handed glove thrown on the ground at de Graeff's feet – a challenge glove, a

throwing down of the gauntlet. Andries de Graeff and Frans Banning Cocq were brothers-in-law. In the Night Watch, Banning Cocq has picked up that right-handed challenge glove. The antagonism had begun." (Greenaway, 2006a, p. VIII)

Fig. 3 - Portrait of Andries de Graeff.

Fig. 4 - Long medium shot of Banning Cocq.

What does Greenaway mean by "antagonism"? The first act of the drama he develops in the screen play is the intrigue of the assassination (plot), the second act reveals the counterplot of unmasking the conspiracy, while the third act is concerned with the revenge of the accused gentlemen. They drive Rembrandt into bankruptcy after he damaged their image with the group portrait later known as The Night Watch. This story of the painter's decline is an integral part of the so called Rembrandt myth, which historians of art have tried to refute since ages. But Greenaway's aim, of course, is not a scientific one. He treats the back story of The Nigh Watch like a crime novel. Mysterious indices are selected and combined to make up the plot pattern (diagram) of motive, murder, and conspiracy, detection, accusation, and revenge. The screenplay displays the results of corollarial and theorematic deductions based on the assumption (or abduction) that a hidden plot is laid out in the group portrait. The movie, then, is a multimedia-performance of this emplotment. Evidently, the only requirement for this emplotment is the relational configuration of shape and color, figures and action on the canvas – the script of roll call and marching order, of aiming and shooting with a musket, of military context and erotic subtext.

Though perhaps many pictures are less dramatic than The Nightwatch, every image is a lay-out of relations and therefore can be considerd as starting point for diagrammatic operations which make use of its de-sign in a hypothetical or speculative manner. What Peirce has called corollarial or theorematic deduction often is not only a logical but also an aesthetic device very closely connected to the display-function of art. Seen this way, it is not too surprising, that Peirce has used the verb 'to perform', surely more an expression of artistic than of scientific origin and relevance, to describe the procedure of theorematic deduction. As already cited above, this way of reasoning "performs an ingenious experiment upon the diagram, and by the observation of the diagram so modified, ascertains the truth of the conclusion." (CP 2.667)

If one compares this description with the ingenious experiment which Greenaway has performed on the diagrammatic structure of the Night Watch, the parallels become evident:

> *"There is a conspiracy painted in Rembrandt's The Night Watch. The sinister title of the painting alone suggests we should look for it. And we should listen too to the sound-track of the painting. Amongst all the hullabaloo, the dogs barking, the drummer drumming, the clattering of thirteen pikes, the hallowing Banning Cocq, the loudest sound is of a musket shot. You can see the flame of the firing, bursting forth behind the head of the foreground figure in yellow, who carries the head of his halberd where the prick should be, and*

whose belly is groped by the shadow of the hand of the companion. Where did the bullet go? We should investigate, and when we do, in the end, with a little genius adventuring, we can plainly see that the whole gaudy endeavor of his painting [...] is going to stir up trouble." (Greenaway, 2006a, cover text)

"To stir up trouble" is exactly what the private eye detective does in the hard boiled novels written by Hammett, Chandler and all the other authors contemplating stories of crime and investigation. The thought experiment performed by Greenaway resembles their work in many respects, especially in that the plot has to be abduced by the author of the novel. The main difference between Greenaway and Hammett, or Chandler is that the British film maker relies on a famous painting while the inventers of the hard boiled novel made no use of this method of palimpsest reading. Beneath the surface of the group portrait and its literal description in Banning Cocq's family album Greenaway detects the story of an assassination. The victim was Pierre van Hasselburgh, the former captain of the company. As a matter of fact, he was shot during a military exercise – apparently an accident, suspiciously a so far undetected crime.

Whether The Night Watch is considered as a group portrait pretending to reveal the reaction to a roll call or as a complex puzzle hinting at an assassination – the precondition is the diagrammatic structure of the image. Its layout can be used as story outline. Complementary, every narrative can be displayed by a series of pictures, film frames or comic panels showing the major plot points. In any case conjecture is necessary. Diagrammatic operations mediate between different genres, arts, and media. Since diagrammatic operations are mental acts, Kant's functional description of the scheme provides a good explanation: As a product of imagination the schema mediates between visual experience (Anschauung) und conceptual understanding (Begriff). The schema allows the spectator to read time (den inneren Sinn der Wahrnehmung) into space (den äußeren Sinn der Wahrnehmung), which is a precondition for the act of emplotment.

But the narrative description (ekphrasis) of a painting seems to be only one type of diagrammatoidal interpretation. Another, probably much more important type is the interpretation of verbal metaphor, where a pictorial concept is projected on a target domain in such a way that relations of correspondence appear and shape the understanding of the sign's object. Of course, metaphorical description is central for Greenaway, since it bridges the gap between painted action and drama, between Rembrandt's group portrait and movie making. He speaks of the "sound-track" – thereby doubling the synaesthetic quality of a picture which illustrates a roll call –

and asks rhetorically: "Is this a painting, an act of theatre, or a still from a film?" (Greenaway, 2006a, cover text)

Based on this suggestive questioning, Greenaway is able to locate his project in a wider frame of reference, including other examples of artful detection.

> *"We intend to make a film about Rembrandt's forensic enquiry in paint, his Crime Scene Investigation [...], and we intend to call our case-history Nightwatching. The fascination of such interpretative investigation of the painted or drawn image goes back for me to the 1982 The Draughtsman's Contract, where indeed a murder was perceived through the contemplation of a set of drawings as they were being made, though the drawings there were 'fictious' and made to contain the plot and not, as here in The Night Watch, the other way around. In film narrative terms, The Draughtsman's Contract has sometimes been compared to the central premise of Antonioni's Blow up, where David Hemmings closely examines the ambiguous details of a photograph that supposedly contains evidence of murder. There is a even more ripe and sinister film – though of completely different nature and import – the Zapruder amateur film shot of the Kennedy assassination where such now evocative emblems as the grassy knoll and the book depository building, frame a landscape, and some perceive, a gunshot flash (not perhaps unlike the gunshot flash in the Night Watch) in a notorious numbered film frame, that certainly did not come from Lee Harvey Oswald's gun and that seriously needs an explanation."*
> (Greenaway, 2006b, p. 3)

Expressions like "forensic enquiry", "crime Investigation", and "case-history" or "premise", and "plot" hint at the strong interplay of logic and aesthetics whenever the imagination performs a thought experiment upon a diagram. Greenaway exploits the interplay of image and drama, pictorial, figurative and narrative signs, already at work in Rembrandt's painting. Because of its complexity The Night Watch has to be read like a text since no one is able to see at first glance what the vast painting displays. One has to start an exploration and one has to find out by observation and conjecture how all the items on the canvas are related and integrated, how they contribute to the overall impression and meaning. Scanning the surface, the observer performs different diagrammatic operations: he reads time into the spatial layout, links several figures and actions, abduces their motives and relates what he sees and explores to what he already knows.

Intentionally, Greenaway's Nightwatching-project is not an illustration of semiotic and diagrammatoidal procedures. In fact, the irony of the project lies in the exaggeration of the conspiracy theory. Greenaway believes that the movie making industry has been exhausted by story telling and has never reached the standard of visual arts. As long as cinema is illustrated novel, it will be degenerate. Movies based on books, scripts, and other text, do not realize the potentials of the medium. To read time into a painting and to transform its relational structure into a story line, therefore, is a misunderstanding of visual art. Seen this way, the whole ingenious confabulation of murder and conspiracy is a reduction ad absurdum. One should not reduce art to back stories and biographism, one should not believe in the Rembrandt myth or any other legendary narrative, namely theories of conspiracy.

Nevertheless, emplotment and confabulation, not too astonishingly, are highly entertaining and inspiring. Greenaway's binary distinction between visual art (containing painting and real cinema) and narrative (containing every type of story telling whether conveyed by books, movies, comics or whatsoever) seems to be misleading, since this distinction neglects the power of diagrammatoidal imagination and understanding. Even abstract formulas like concept art and action painting display the story of their genesis and effects. Whether one likes or dislikes emplotment and confabulation, it is hard (if not impossible) to make a drawing or a painting, an image, a still from a feature film, or any other kind of audiovisual art without relational structure. No one can escape the display function of art or the mechanism of conjecture and projection. Far from de-constructing the predominance of text, narrative and history, Greenaway, starting with an ekphrasis, unfolds the tellability of image and imagination, links the outline of Rembrandt's group portrait with the genre formula of the detective novel and treats every figure, gesture, and relation as hint, clue, or index. By devising a conspiracy theory to explain the riddles of The Night Watch Greenaway performs an abduction. In this frame of reference, his screen play progresses by theorematic deductions which are controlled inductively since every hypothetical emplotment and confabulation is backed by the painting and its context. If, for example, Banning Cocq carries a right glove in his gloved right hand, the riddle is solved by referring to another painting. The relation that can be observed between The Night Watch and the Portrait of Andres de Graeff makes evident the diagrammatoidal feature of Greenaway project. The logic of his emplotment and confabulation resembles the logic of constructing a diagram and performing a thought experiment – according to Peirce the main features of diagrammatoidal reasoning:

> *"By diagrammatic reasoning, I mean reasoning which constructs a diagram according to a precept expressed in general terms, performs experiments upon the diagram, notes their results, [...] This was a discovery of no little importance, showing, as it does, that all knowledge without exception comes from observation."* (NEM, 1976, IV, 47)

Peirce himself has illustrated the concept of creative, abductive thinking and diagrammatoidal reasoning with respect to Johannes Kepler's exploration of the elliptical shape of planet movement. After the perfect, godlike circle did not fit the data, Kepler was forced to take into account the derivate shape with two foci. Surprisingly, this figure suits to the facts much better and provides a schematic tool to deduce further insights about astronomy. Kepler tried – at first unsuccessfully – to solve a scientific problem by testing aesthetic figures as schemata for observation and conjecture, theorematic deduction and empirical induction.

In fact, the exact shape of the Mars orbit is much more complex than its comparison with an ellipse suggests. Nevertheless, Peirce's logic of exploration is convincing: all knowledge (insight) comes from observation. Since the term 'observation' does not imply that its object belongs to the material or the visible world, imagination can (and often must) enter the aesthetic process of establishing a schema or a theory to explain facts which otherwise would remain mysterious. Art, then, is a knowledge producing activity, or, to put it the other way around: thinking without imagination is not able to produce any new idea.

But there is another point worth mentioning. A thought experiment, by definition, fulfills its task in the imagination only, while every work of art has to use some material in some way. One might say, that thought experiments can be written down and that they sometimes are developed through narration (as is the case with Greenaway's Nightwatching project,), and narration, surely, is a certain type of materialization. Even concept art needs a display that transcends the sphere of pure mental activity.

Since every idea or concept has to be incorporated in some material, the display function of art often is realized through a specific form of de-sign which Nelson Goodman has named "exemplification". Exemplification is defined as "reference based on possession" (Goodman, 1995, p. 60) and in this respect closely connected to Sybille Krämer's idea of "incorporation" (Krämer, 2006, p. 83). At first glance, it seems that exemplification is a kind of indexical representation. Nevertheless, it might enrich our understanding of iconic representation also because an image or a diagram can incorporate the "Gestalt" or the "Schema" of the object it refers to.

silence silence silence silence silence
silence silence silence silence silence
silence silence   silence silence
silence silence silence silence silence
silence silence silence silence silence

The blank in the middle shows what this poem, originally invented by Eugen Gomringer, is about. The text incorporates the object it refers to and makes visible what no one can see. It seems bizarre and, of course, it is paradox, but while the word "silence" does not possess what it means, the poetic configuration of letters around a void does 'incorporate', exemplify, or display the meaning of 'silence'. Apparently, exemplification bridges the gap between abduction, theorematic deduction and induction because it is a matter of imagination and observation.

3. Example: Oliver Stone's JFK

There are several genres that blend together exemplification and investigation: detective stories, police procedurals, and the movies showing a court proceeding or a trial. Oliver Stone's JFK (USA 1992) unites them and makes great use of the collaboration of exemplification, perceptual judgments, and diagrammatoidal thinking – thereby showing the capacity of scenographic media to give evidence to dicisigns and arguments as used in the rhetoric of charge and defence. JFK is based on Jim Garrison's recollections of his fight against the Warren Commission's report on the Kennedy assassination in 1963. Garrison's attempt is crucial to refute the theory of the magic bullet that should explain why so many different wounds could have been caused by a single shot. Garrisons wants to convince the public that several shooters were aiming and firing at the president. If so, a conspiracy was going on.

Oliver Stone's movie is a brilliant conjecture of documentary and feature film material. The set up for Garrison's speech is complicated by a lot of inserts. While Kevin Kostner delivers the speech, the Zapruder amateur film, authentic and inauthentic photographs, a city model and body sketches are presented as indices. The actor's speech, then, is much more than a talk: it is also the public performance of a thought experiment: What must be inferred from the facts, if the theory of the magic bullet is wrong? Garrison investigates and, with a little help of diagrammatic montage, comes up with the conclusion that Lee Harvey Oswald was a kind of spacegoat, killed before he could reveal the true story about the Kennedy assassination.

A closer look at the key sequence shows that Stone himself has manipulated some facts to make evident the conspiracy theory. For example, he confuses the right and left hand of the politician who was sitting in front of Kennedy. Nevertheless, the court sequence in JFK is an overwhelming exemplification of diagrammatoidal reasoning. The interplay of verbal speech and visual presentation, moving images and diagrams, observation and imagination, abduction, deduction and induction exhibits both the method of diagrammatoidal reasoning and the display function of art. Every requisite, gesture and speech act, every shot and frame functions as dicisign, while the whole sequence exposes the rhetorical force and strength of scenographic media. One might even say, that a derivate of the Pragmatic Maxim rules the representation of events and arguments: Consider what effects, which might conceivably have lethal consequences, you can conceive of a magic bullet and compare this highly implausible theory to the practical bearings, you can conceive of, if the hypothesis of conspiracy is taken into account seriously. Then the whole of the diagrammatic reasoning is our idea of the hidden plot of the Kennedy assassination.

Fig.5 - Mastershot of the Garrison talk.

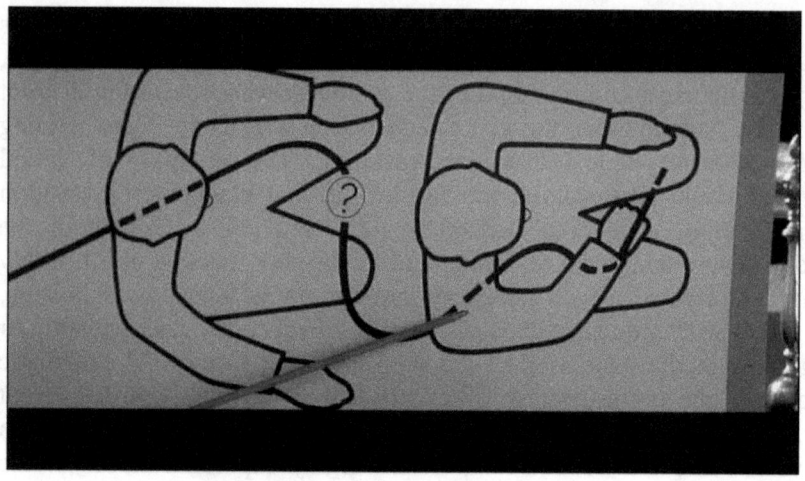

Fig. 6 - Diagram of the magic bullet's journey
through time and space.

Summary

The basic fact is that scenographic media can perform and display diagrammatic operations. In a virtual space they make abductions evident and (theorematic) deductions observable. Exemplification is an artistic way to transform the Pragmatic Maxim into a simulacrum and therefore functions similar to induction. In short: scenographic media make thought experiments visible by performance and exemplification.

As Greenaway's Nightwatching project shows, diagrammatic operations are unavoidable to re- or de-construct a topic or a plot and to transform it from one medium/discourse to another. There are at least three operations:

to discriminate between performance and topic/plot
to display the relational structure of the topic/plot
to perform this structure in compliance with the target medium/discourse

It is of crucial importance, that all three operations involve imagination and contribute to a creative process that combines observation and reflection.

Furthermore, the media join this process in two ways. On the one hand, the creative process starts with one medium (or: with one configuration established in compliance with a certain medium) and leads to another medium (or: configuration). On the other hand, additional media may catalyze the process of display.

If for example a feature film includes frames which mirror a certain painting (The Night Watch), there is not only the feature film and the painting. Rather a photograph of the painting, a script or a story board play intermediary roles in the process of re- or de-constructing the painting in the movie.

While Greenaway's Nightwatching-project deconstructs the myth connected to Rembrandt's bankruptcy in a very autoreflexive manner, showing the pitfalls of biographism etc., Oliver Stone reconstructs the scrutinizing of the Kennedy assassination, using films, stills, maps, diagrams, performances, and, of course, testimonies.

In both cases, scenographic media display a thought experiment and, since performance needs incorporation, visualize abstract relations and operations in a dramatic manner.

Since Stone quotes forensic investigations and Greenaway exploits the seminal work of scholars and historians, their scenographic attempts show how diagrams are used and how they stimulate and shape the interplay of observation and imagination according to Peirce's trichonomy of abduction, deduction, and induction.

Therefore the logic and aesthetic usage of diagrams and diagrammatic operations do not differ too much. Scientists and artists abduce diagrams to deduce something and then test corollarial or theorematic conjectures by induction. The display-function of art is a way to give evidence to everything that can be inferred from a given plot, scheme, or pattern. Similar to an architect who explores possibilities by drawing a sketch, a lot of world making starts with a virtual test by incorporating possibilities in such a way that one can discriminate the working possibilities from the not working ones.

References

Bauer, M. (2005), Schwerkraft und Leichtsinn. Kreative Zeichenhandlungen im intermediären Feld von Wissenschaft und Literatur. Freiburg.

Goodman, N.: (1995), Sprachen der Kunst. Entwurf einer Symboltheorie. Übersetzt von Bernd Philippi. Frankfurt am Main.

Greenaway, P. (2006a), Nightwatching. A View of Rembrandt's "The Night Watch". Amsterdam.

Greenaway, p. (2006b), Nightwatching. [Screenplay]. Paris Barcelona.

Krämer, S. (2006), "Die Schrift als Hybrid aus Sprache und Bild. Thesen über die Schriftbildlichkeit unter Berücksichtigung von Diagrammatik und Kartographie. In: Hoffmann, T. et. al. (eds.), Bilder. Ein (neues) Leitmedium? Göttingen, p. 79-92.

Peirce, C. S. (CP), Collected Papers. Vol I-VI ed. by. C. Hartshorne et al. Cambridige 1931-1955; Vol VII-VIII ed. by. A. W. Burks, Cambridge 1958.

Peirce, C. S. (NEM 1976), New Elements of Mathematics, ed. by. C Eisele, The Hague.

Semiotics and Sgnifics (1977). The Correspondence between Charles S. Peirce and Victoria Lady Welby, ed. by C. S. Hardwick. Blomington London.

III. On Maps

The World in Maps:
Looking for Treasures, Neurons, and Soldiers

by
Valeria Giardino

Abstract

In this article, I will discuss the use of maps, beginning with geographical maps and arriving at new cases of maps such as cortical maps and graphs. The main idea of the article is that to use a map there are different components that must be considered. First, a map must display a space that is organized in different shapes. Then, more crucially, it must be correctly interpreted. Moreover, there is a sense in which a map is always internally and externally incomplete. Finally, a map is precise, because it can be inaccurate and lie but it will always represent a particular space of possibilities. I will show how these components transform when applied to the new kinds of maps.

1. Introduction: there is more than just a map

"I know by the color. We're right over Illinois yet. And you can see for yourself that Indiana ain't in sight."
"I wonder what's the matter with you, Huck. You know it by the COLOR?"
"Yes, of course I do."
"What's the color got to do with it?"
"It's got everything to do with it. Illinois is green, Indiana is pink. You show me any pink down here, if you can. No, sir; it's green."
"Indiana PINK? Why, what a lie!"
"It ain't no lie; I've seen it on the map, and it's pink."
- Mark Twain, Tom Sawyer abroad

Huck Finn and Tom Sawyer are flying over the United States of America in a balloon. Huck looks down and asks himself why, though they should already be beyond Illinois, he cannot see Indiana. Toms asks him why he is so sure about that: they are miles away over the countryside, in a balloon, there is no way to know that it is not yet Indiana down there.
Nevertheless, his friend has no doubt. Of course they are not yet over Indiana, he can see it from the color: Illinois is green, while Indiana is

pink. From above, there no rose land that can be seen. When Tom blames him because he thinks he is lying, Huck answers that he is not lying at all, because he has seen that Indiana is pink on the map, and if maps are supposed to provide us with facts, then it is a fact that Indiana is pink.

Huck is right in saying that the color of Indiana in the geographic map is pink. Nonetheless, he makes an error when he thinks that because of that, Indiana is also pink in reality. Evidently, the boy is not familiar with maps, and as a consequence he does not know that they have intrinsic expressive limitations. In particular, Huck does not understand that the map is colored, but not for the reason that he naively thinks. Indeed, colors on the map do not correspond to colors in reality, but they have the primarily anthropocentric function of sharply marking the separation between two different regions of interest. Since the color of a region separates that region from the neighboring ones, the choice of the color of two adjacent regions is subject to the visual objective of making this boundary standing out. Indiana and Illinois on the map could surely be violet and lilac instead that being pink and green, but then this objective would be more difficult to meet. Of course, it is because of our nature that some colors appear to us more sharply distinct than others. We can suppose that a different species that sees different colors would maybe color maps in a different way. Nevertheless, the expert user, and even an intelligent young boy such as Tom Sawyer, knows well that in reality there is no land that is pink, in the same way in which borders in reality do not correspond to any black track that has been dug into the earth.

So, what is a geographical map in the end? A map (in particular a geographical map) is a tool that the observer uses in order to find his way in a particular region of the space, a city for example, and in order to choose the right path that will bring her to her final destination. To be able to use the map, there are pieces of knowledge not directly present in the map that the observer must be add to it. First, she must identify the map as a map; in other words, she has to identify the map as a model of reality, and not as reality itself. Second, she must be able to put into correspondence the spots on the map with the different places in the city. Third, she must also know that she can be inside the map, that means that the map can contain her, as it happens when, for example in a museum, there are maps with a red dot that indicates "you are here". Combining together these pieces of information, which are extrinsic to the map, she can finally move in the space of the city.

One possibility is to think that a map recalls the space it represents because it is made in such a way that it preserves, in a minor scale of course, the same spatial relationships that are present in reality. Nevertheless, such notion of "resemblance" is dangerous, because maps are not exactly icons that "resemble" what they represent: the map user must rather learn to read them, and conversely she must also learn to read reality by means of maps. Maps live in fact in the tension that creates between their spatial dimension, which makes them as real as any other objects in space and time, and their right interpretation, which the user must know to correlate spots on the map and places in the world. Why is Huck wrong in thinking that Indiana seen from above is pink? Neither reality is, as he believes, a faithful reproduction of the map, nor the map is, as it is evident, a faithful reproduction of reality.

In this article, in section 2 I will analyze the use of geographical maps, and I will identify four components that they must possess in order to be used to orientate in the space. In section 3, I will present new forms of maps, and discuss whether they still possess the four components that are typical of geographical maps. In section 4, I will draw some conclusion and claim that when we are appropriately using a map, we are referring to more than just a map as a physical external object.

2. How to use a map

2.1. Territories to describe

Cognitive anthropology teaches us that geometrical intuitions based on points, lines, parallel lines, right angles, and so on, are universal to all human beings, despite their degree of education; for this reason, they are legitimately part of our cognitive core abilities.

The Munduruku, a native population in the Amazonian forest, do not possess a very sophisticated language. Nevertheless, they are able to use maps: they show no difficulty in connecting simple shapes on a piece of paper to different places in space. As a consequence, they succeed in tasks such as tracking objects that are on the map but cannot be found in the space, and conversely in tasks as finding objects which are hidden in space but whose presence is shown on the map[1]. The results of this study is thus in line with the idea that it is possible to have elementary geometrical intuitions also without having received a specific education; furthermore, they seem to suggest that this happens even without possessing a language that allows for the use of geometrical terms. Therefore, human beings would spontaneously see shapes without necessarily giving a name to them, and moreover they would be able to know what to do with them, creating correspondences between these shapes and the world around them.

Nevertheless, it is a difficult enterprise to draw a sharp distinction between purely perceptual capacities and purely linguistic ones. In fact, the

---

[1] Dehaene et al. (2006)

problem of finding a clear definition of the role that the acquisition of language plays in our cognition of the world constitutes a challenge for the most recent psychological approaches, which at the moment do not seem to have provided any definitive theory. This does not mean that it is not possible to find any distinctive feature that characterizes maps; as we have seen, the most crucial and simple component for a map to be a map, is its

(i) Spatiality: a map describes a space.

What a map does is to display more or less complex spatial relations among the elements it contains and which the map recipient must be able to track. The shapes that are recognized in and on the map are at the base of the construction of the correspondences with reality. The user looks on the map to understand what she should expect to find in reality: the nature of this correspondence, differently from what would happen with a linguistic description of the same place, calls for an analogy between the organization of the space inside the map and the organization of the space outside the map. In fact, the map becomes a key to reality, a way of giving an order to it.

Another aspect to point out here is that a map is also holistic. In fact, rather than considering all its elements in isolation, the user must see the global structure of the relationships that it displays, and within which the local details must be accounted for. In the following sections of this article, I will show the way in which because of their spatiality, maps will transform in other kinds of objects: continuous space offers itself as an effective tool also for conveying information that is relative to more complex relationships than the simple distance between places, such as for example the development of a process through time.

Several studies on the geographical explorations in the New World in the fifteenth and sixteenth century have tried to reconstruct the possible exchanges that must have taken place between Europeans and the native population, precisely to the aim of creating maps that would define the territory left to the conquerors to be explored[2]. There is no doubt that guides and native interpreters must have provided important explanations and descriptions, whenever it is difficult to evaluate in detail to what extent; nevertheless, to clarify the role of spatiality in maps, it is interesting to consider what Bartolomé de Las Casas tells us in his History of the Indies. He reports that on an occasion some natives, who have been brought to Europe as prisoners, showed that, when asked, they could create maps. According to his story, the King of Portugal John II, meeting Cristoforo Colombo, ordered to their servants to bring some beans in the hall, and asked one of the natives who were present to use the beans to draw a map of the

---

[2] Harley, J. B. (1992)

archipelago of islands where he was born. The native, very naturally and without showing any effort, displayed the beans on the table, and used them to represent the islands of Hispaniola, Cuba, the Lucayas, and the other territories that he knew. The King then asked another native to display the lands in that sea, and the second native did not show any difficulty in recognizing an accurate map in the disposition of the beans proposed by his companion. What he did was grasping other beans and putting them near to the previous ones, explaining to the Europeans in his incomprehensible language that he had added some more information[3]. Therefore, the natives showed to possess the cognitive abilities that are necessary to understand that the dislocation of the beans can be used to express the spatial relations that are exhibited by the archipelago.

Nevertheless, these simple activations of spatial recognition cannot be considered in isolation with respect to other higher-level capacities that are in play when using a map. Beginning with the shapes that have been selected in our visual system, we must establish a series of correspondences in a map if we want to use it: the function of the map is to lead us in the discovery of the represented reality. In a recent work, Elisabeth Spelke and colleagues has shown that young children are capable of extracting and employing the geometrical information that is present in a map without even recurring to instructions or explanations given by adults, as Munduruku do[4]. This capacity does not appear before 4 years: only at that age the child is able to track geometrical correspondences between tri-dimensional world and two-dimensional maps, using the extracted information to orient herself and to move in the space. This age - 4 years old - seems to be crucial in the cognitive development of the child, since it is at the same age that she also develops her arithmetical capacities.

2.2. Why every map is a treasure map

In Robert Louis Stevenson's novel, Treasure Island, the young Jim Hawkins finds in the chest of Captain Billy Bones, died in mysterious circumstances in Jim's parents' inn, a map that he is not able to decipher. To understand what the map is about, Jim brings it to her friends Trelawney and Livesey, and the two, very excited, recognize in it the treasure map of Captain Flint, a pirate known for his cruelty and disappeared without having left behind him any trace of his enormous possessions.

---

[3] Parry, J. H. and Keith, R. C. (1984)
[4] Shusterman *et al.* (2008)

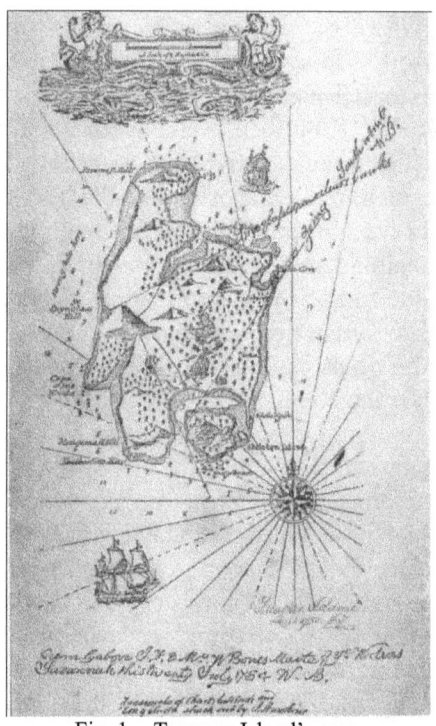
Fig. 1 - Treasure Island's map

The map that Jim has found is a map:

> *"with latitude and longitude, soundings, names of hills and bays and inlets, and every particular that would be needed to bring a ship to a safe anchorage upon its shores. It was about nine miles long and five across, shaped, you might say, like a fat dragon standing up, and had two fine land-locked harbours, and a hill in the centre part marked "The Spy-glass." There were several additions of a later date, but above all, three crosses of red ink--two on the north part of the island, one in the southwest--and beside this last, in the same red ink, and in a small, neat hand, very different from the captain's tottery characters, these words: "Bulk of treasure here.""*[5]

On the reverse side, with the same orthography, there are also other indications on how to reach the island, and on where the treasure is to be found. For the protagonist, the map, though it has a shape that resembles that of a dragon, is incomprehensible. In fact, Jim knows that it is a map, and he

---

[5] Stevenson, R. L. (1883), from Chapter 6.

also understands that the cross on it must represent the place where the treasure is hidden, as the words indicate as well; nevertheless, he does not have the means to read it. In fact, neither he knows where Treasure Island is, nor he is sure that it exists at all. Yet, he has no difficulties in recognizing spatial relations on the map. What he does not know is whether he is entitled to think that the things depicted on the map really exist.

We could think that when we use an ordinary map we do not have the same problem Jim is having, since we are surely not looking for a pirate treasure or looking for a mysterious island somewhere out there in the ocean. Nevertheless, this is not completely true. To some extent, every map always talks about a treasure we want to find: this treasure is the place we want to go to. To reach our objective, we cannot just look at the map, but we have to possess some other elements. Actually, we have to know:

that the map is a map of some place;
to which places the locations on the map correspond in reality;
that we can put ourselves inside the map;
that we can choose a destination on the map.

This knowledge is neither already present in the map, nor it is derivable from it. In this sense, every map ends up in becoming a treasure map, since to be read, it involves each time other pieces of knowledge that we are demanded to possess already.

A second component of maps is therefore the following:

(ii) Interpretation: a map needs to be read, not only seen.
A bad interpretation of the map translates in a bad reading of reality: Huck reads the map literally because he ignores a shared convention, and for this reason a piece of information is lost.

2.3. Incomplete maps

Suppose now we look at a map of a city that we have never visited but that we know because we have read its history in a book. How would this map be analogous to the linguistic description of the city?

We know that a description, to be effective, does not have to be exhaustive. If the observer simply reports one by one all the elements which are present in the scene, the description provided would not be interesting at all and most of all it would be nothing like what we ordinarily consider a description to be. A description is rather the product of the selection of what is relevant in the scene on the one hand, and of the omission of what is not relevant in it on the other. Therefore, also a description results from an interpretation.

In the case of the city, we can for example decide from time to time in which way we want to 'see' it. That would imply that each time a different description will be provided. We could for example forget about the color of the building façades and linger on their majesty, or leave out the orientation north-south of the streets and classify the trees that we see. In her activity of describing what she sees, the observer does not want so much to report the greatest possible number of elements in the visual field; her objective is rather the disclosing of the relationships she sees among these elements. In this respect, a map is not different from a description, since not all reality may correspond to what it conveys.

Jorge Luis Borges appreciates this feature of maps when he tells us about an Empire of the past where the art of cartography had reached such a perfection that

> *"... the Map of a Single province covered the space of an entire City, and the Map of the Empire itself an entire Province. In the course of Time, these Extensive maps were found somehow wanting, and so the College of Cartographers evolved a Map of the Empire that was of the same Scale as the Empire and that coincided with it point for point."*[6]

A map as the one described here by Borges, which is so wide that it coincides point by point with reality, is maybe perfect but it has a defect: it is completely useless. In fact, Borges tells us, the succeeding generations judged that a map of such magnitude was "cumbersome", and abandoned it to the rigors of sun and rain. Here Borges plays with the words: from the rigor of science it wanted to reach, the map is now left with the rigor of the weather.

---

[6] *Of Rigor in Science*, quoted from Borges (1975). Something similar happens in one of Calvino's *Invisible Cities*, the city of Valdrada, which is built on the shores of a lake, with houses which are all verandas one above the other. "Thus the traveler, arriving, sees two cities: one erect above the lake, and the other reflected, upside down. Nothing exists or happens in the one Valdrada that the other Valdrada does not repeat, because the city was so constructed that its every point would be reflected in its mirror, and the Valdrada down in the water contains not only all the flutings and juttings of the facades that rise above the lake, but also the rooms' interiors with ceilings and floors, the perspective of the halls, the mirrors of the wardrobes." See for reference Calvino (1972).

For these reasons, a map, exactly as a linguistic description, is not a reflection of reality, and a third component of maps emerges. I will call this third component "internal incompleteness":

(iii)   Internal incompleteness: a map is never exhaustive.

Nevertheless, there are also other crucial respects in which a map seems to resemble more an image than a description. The map is in fact a space of possibilities, for the reason that it asks its user to figure out the places it represents and to imagine that she can move from one point to the other by means of it. The observer, beginning with the simple signs traced on the paper and by looking at the map, sees before her a whole world of possibilities, of choices, of travels and of possible returns. It is therefore true that even if on the one hand, as Borges shows us, a map must be incomplete to be effective, since it is not supposed to be a faithful reproduction of reality, on the other hand it is necessary to avoid the opposite risk: reality cannot become a faithful reproduction of the map, in such a way that it becomes possible even to travel on it. That is what happens in Cervantes' Don Quixote, where the author explains to the reader how easy it is to travel on a map compared to travel in reality: on a map, the traveler does not have to face any effort, because he is never thirsty nor hungry, never cold nor hot[7].

Another typical component of maps emerges here. I will call it "external incompleteness".

(iv)   External incompleteness: a map is a space of possibility.

To travel on a map and explore the world of possibilities that it discloses, the observer recalls past experiences and memories. The content of the map is in the end potentially wider and more complex than the object that the map represents, since it is a function of all the possibilities that are offered to the observer. The observer sees the map as she desires it to be, and she even sees herself as she desires her to be inside the map: for this reason, the map has a fascinating power on her.

Among the cities dreamt by Calvino in his Invisible Cities, there is the sophisticated city of Eudoxia, which exemplifies this idea. At Eudoxia, a carpet is kept where the shape of the city is shown. At first sight, nothing in the drawing of the carpet arranged in symmetrical figures that repeat themselves along circular and right lines seems to resemble to Eudoxia. Nevertheless, if the inhabitants observe the carpet more deeply, they are persuaded that each area in the drawing of the carpet corresponds in fact to a place in Eudoxia, in such a way that the itself city is contained in the carpet

---

[7] On this aspect of maps, see Muehrcke and Muehrcke (1974).

as well: it is in the carpet that the all the relations in Eudoxia get fixed. Here we see maps' external incompleteness:

> *"Every inhabitant of Eudoxia compares the carpet's immobile order with his own image of the city, an anguish of his own, and each can find, concealed among the arabesques, an answer, the story of his life, the twists of fate."*[8]

The relationship between the city and the map is mysterious. The inhabitants decide to question an oracle about that. The oracle replies that one of the two objects has the form the gods have given the sky and the orbits in which the worlds revolve; the other is an approximate reflection, like every human creation. The doubt remains on which of Eudoxia and the carpet is real and which is only a map that represents reality: what would prevent us to think that Eudoxia, "a stain that spreads out shapelessly, with crooked streets, houses that crumble one upon the other amid clouds of dust, fires, screams in the darkness", is not itself a map of the universe?

On the one hand there is chaotic reality when we can get lost, on the other hand too much abstraction can go too far and transforms into self-reference: maps seem to be in equilibrium between the pole of chaotic reality and the one of the excess of abstraction.

2.4. A map may lie but it never jokes

There is also another aspect of maps that has to be considered when reading or drawing a map, which is its natural degree of reliability. We can doubt the existence of what a map represents, as in Stevenson's case, but we cannot doubt the fact that if this place exists, it would be exactly as the map represents it. As Huck keeps repeating, maps are not done to anything else but to give access to facts.

Maps create the world they represent in the very same moment they represent it: they can lie on its existence, but by offering themselves as maps they cannot avoid describing it for what it is, even if they can still reveal themselves to be not accurate enough. Therefore, a map can lie, but by doing that it nonetheless takes seriously the object to which it refers, and for this

---

[8] Calvino, *ibid.*

reason it can never joke[9]; to draw a false map is nonetheless a kind of imitation of reality[10].

Here we find the fifth and last component of maps:

(v) Precision: a map searches for the truth.

At this point, we can sum up the different aspects that characterize the nature of geographical maps:

spatiality
interpretation
internal incompleteness
external incompleteness
precision

In the next section, I will argue that new images and new diagrams develop from maps. I will try to evaluate how the aspects (i)-(v) are present in them and in which degree.

3. New maps

3.1. Cortical maps

In the creation of the maps I will consider now, technology plays a considerable role. Cortical maps are the classic example. These maps were born with the discipline of phrenology, which was developed by the German physician Franz Joseph Gall in the middle of the 19th century[11].

---

[9] "A map may lie, but it never jokes" is a line of the poem "Listening to Maps" by Howard McCord, taken from his collection McCord

[10] This feature can be used to manipulate the public opinion, as for examples in the case of maps used in advertising or in politic propaganda. There are many interesting exemples in Monmonier, M. (1991).

[11] Gall (1819)

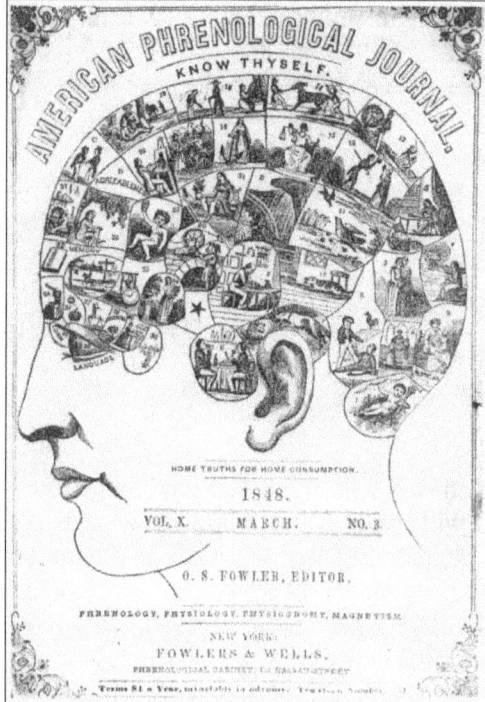

Fig. 2 The brain as represented in
The American Phrenology Journal in 1848.

Phrenology was based on two assumptions. The first assumption was the idea that the brain corresponds to the mind, and therefore it is possible to find different areas inside the brain; each of these areas is supposed to be dedicated to one particular faculty or to one particular feeling. The second assumption was that these areas, since they were localized on the surface of the brain, pressed on the skull, and that therefore it was sufficient to analyze the conformation of the skull of an individual to predict his personality, his character, his moral inclinations and his intellectual capacities.

If the second assumption has been totally refuted by contemporary neurophysiology, the first assumption has someway survived trough last century until the relatively recent development of "brain-imagining" methods, such as the PET (Positron Emission Tomography) and the fMRI (Functional Magnetic Resonance Imaging), thanks to which it has become possible to register brain activity during the processing of some perceptual, attentional, emotional behaviors. Using these tools, subjects are tested in a non invasive way with the final aim of discovering which cerebral structures are dedicated to particular forms of information treating. Brain-imaging utilizes already known proprieties of cerebral metabolism, as the correlation between neuronal activity and blood influx, and is based on advanced techniques of detection. The study of the brain in relation to the mind and together with it the models of cognitive psychology has finally found a territory when they can be put to the test: the creation of new anatomical and at the same time functional maps of the human brain.

It is true that twentieth century experimental cognitive psychology is very distant from phrenology: the ambition of today is basically to find the transition rules that allow for the correlation between the activation of receptors and synapses, between synapses and the activation of neurons, between neurons and cortical columns, cortical columns and regions and circuits of neurons, and finally between populations of neurons and behaviors, culture and education.

The case of these new cortical maps actually shows how it is possible to strongly rely on (i), the spatiality of the cortical map, without taking into account the role and the importance of (ii), interpretation. Despite the flourishing of these experimental investigations, one question to ask is in fact whether these works come together with careful considerations on their methodological set up: PET and fMRI are so versatile as techniques that in some cases they can not only be used but also abused. Without considering the more technical aspects related to accuracy and to the limits of resolution that the available methods allow at the moment, it is necessary to point out that behind the application of these techniques, there are some crucial theoretical presuppositions. First, it is assumed that by measuring metabolic variations in particular experimental contexts, it is possible to demonstrate that there exist populations of neurons that are specific to particular tasks; secondly, it is assumed that it is indeed possible to isolate the activation of neurons in relation to a specific behavior.

In cognitive neuroscience, the researchers often use images of the brain to represent brain activation. These brain images are framed in an undifferentiated grey background and some of their areas are colored, to the aim of indicating their activation. Scientists and experts have suggested that these very peculiar images confer a high scientific credibility to the studies on cognition, and constitute one of the main reasons why the big audience shows itself to be very interested in the researches that make use of brain imagining. The enthusiasm around these works, nevertheless, casts some doubts in the scientific community: the persons involved in the scientific work are worried because of the tendency in the popularization proposed, in particular by the mass media, towards the oversimplification if not towards a real misunderstanding of some of the conclusions. Some people have defined this phenomenon as neurorealism, since it transforms some theoretical choices in a-critical stances, which are yet considered as reliable[12].

In a recent study, the credibility level of the results in cognitive neuroscience has been investigated, to the aim of evaluating if really brain images could have some persuasive power[13]. The tendency to assign a notable degree of credibility to images could be related to our natural disposition at explaining in a reductionist way every cognitive phenomenon. We demand science to tell us which area of the brain would correspond to a particular behavior: we prefer physical representations such as the brain images introduced above to other more abstract representation, such as tables or graphs, for the very reason that they are made of the same elements that we suppose the phenomenon is made of.

In one of the experiments, the researchers have asked to the subjects to carefully read some articles in which the experimental results were shown either by means of topographical maps of brain activation or by means of brain images. The topographic map and the brain image are analogous, since in both cases we have complex images that represent brain activation by means of different colors. The subjects were then asked which of the arguments were correct.

The results are not surprising: the works accompanied by the brain images were considered more credible from a scientific point of view than the ones accompanied by the topographical map. This effect was present also in the case of fictional articles, that means of articles that were made up on purpose by the experimenters and that contained sensible measurement errors. This experiment shows that it makes sense to believe that brain images are more persuasive. One possible explanation is exactly that brain images are images that can easily be interpreted as simple depictions than any other more abstract forms of mapping.

If this is true, then it is interesting to point out that the brain images so obtained, though they give the impression that they are a direct measure of the physical substrate of cognitive processes because they represent a brain that literally "lights up", are actually the product of very complex theoretical strategies and assumptions. These images are based on and are made readable by functional magnetic resonance imaging, which only measures the relative oxygenation of blood in some specific areas of the brain. Therefore, the very functional magnetic resonance imaging method gives the measure of brain activation only indirectly. For this reason, brain images of this kind, whatever they may seem, are functionally analogous to other cortical maps, and do not correspond at all to a direct depiction of

---

[12] Racine *et al.* (2005)
[13] McCabe and Castel (2008)

mental states; in the end, these images are an interpretation of brain activation in relationship to some particular behaviors.

3.2. Maps evolve

Around the seventeenth century, a new genre of maps appears on the scene. These new maps combine cartographic abilities and statistical knowledge and become "data maps". For the first time, not less than 5000 years after the creation of the first geographical chart on clay tablets and many years after the subsequent appearance of many sophisticated geographical maps, cartography begins to collaborate with other disciplines such as statistics[14]. It will be necessary to wait until the end of the nineteenth century to find that the spatiality of the map is eventually used to represent a variety interrelated phenomena or a series of abstract processes. The new graphs offer themselves as more effective tools than lists of propositions or tables, for the very reason that they show the "shape" of the data. In this way, it becomes easier for the observer to make comparisons directly on the image that is displayed before her, without relying on calculations.

A superb example of a data map is the table that illustrates Napoleon's Russian campaign in 1812. This table has not been given at the time Napoleon was invading Russia, but fifty years after the end of the war, by Charles Joseph Minard, who was interested in creating techniques that would innovate the methods of information displaying. Minard's idea was very simple: he wanted to present all the information about one military event - the defeat of Napoleon's army, in a single display. To show the difference between this map and a description in words, remarkable as it is, of the very same defeat, I will quote here Tolstoj's War and Piece:
"From Moscow to Vyazma the French army of seventy-three thousand man not reckoning the Guards (who did nothing during the whole war but pillage) was reduced to thirty-six thousand, though not more than five thousand had fallen in battle. From this beginning the succeeding terms of the progression could be determined mathematically. The French army melted away and perished at the same rate from Moscow to Vyazma, from Vyazma to Smolensk, from Smolensk to the Berezina, and from the Berezina to Vilna-independently of the greater or lesser intensity of the cold, the pursuit, the barring of the way, or any other particular conditions. Beyond Vyazma the French army instead of moving in three columns huddled together into one mass, and so went on to the end."

---

[14] For the history of "data graphics", see for reference Tufte (1983)

There is a reason why Minard's table in Fig. 4 has been considered as 'the best statistical graph ever created'. The observer looks at it, and before her she has many pieces of information of different sorts in one single display. First, the table is a familiar object because it is a map. We can see in it the rivers that the army has crossed, the exact distances that have been covered, the name of the places and of the main battles.

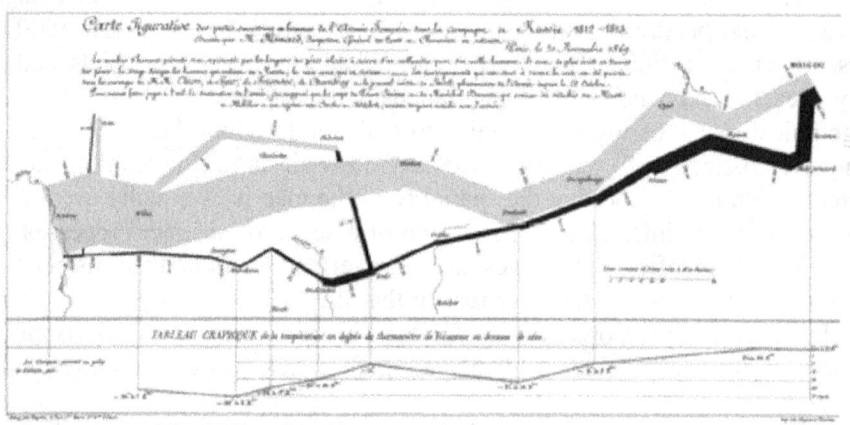

Fig. 3 Minard's table on Napoleon's Russian campaign.

The light grey area represents Napoleon army advancing; the black area, which is narrower, represents the same army withdrawing. The two areas get narrower and narrower because their width is proportional to the number of soldiers that are crossing Russia: each millimeter corresponds to 10.000 men. Other data are contained in the map: below, we find the time that the crossing has taken and the temperatures registered during winter. Some labels underlines the name of the rivers, some other the precise number of soldiers. The presentation of this information directly on the map has the function of putting the observer in the condition of following the displacement of the army in relation to a variety of parameters: the observer reads all this information in the same space of the figure, and for this reason the figure promotes inference, comparisons and projections.

Minard's map tells the military history of the French defeat. Of 422.000 soldiers who left for Poland in 1812, only 100.000 will reach Moscow. During the withdrawal, the percentage of the dead is even greater. A remarkable and sudden narrowing of the black withdrawal area is seen on the river Berezina, where Russians attacked the French soldiers while they were crossing the river, and many of them got killed. In the end, only 10.000 soldiers will come back to France.

The table is interesting because it shows how the creation of maps has now freed itself from its privileged relationship with the geography of the world: the fact that the geographical element is not anymore central allows for a greater abstraction and for a greater richness of content. Graphs tell us today about public debt, demography, steel production, epidemiology, and so on. Maps have evolved into new sorts of formats by means of which the observer aims at reaching new objectives.

We can now ask whether also for graphs of this sort the map components (i)-(iii) defined in the previous section apply. Actually, also graphs use shape to convey information, and, like maps, they are also incomplete, since they are the product of precise choices on which variables to favor among others. For what regards components (iv) and (v), in the case of graphs external incompleteness should be at the minimum, and conversely precision should be at the maximum; if this does not happen, then the graph that shows quantitative data falls victim of distortions. For this reason, there are two possible risks in drawing graphs inadequately: inserting perceptual deformations and lacking expertise and relevant knowledge. Contrary to maps, a graph can lie and it can also joke, with the result of mystifying, as in a caricature, the image of the data that it is supposed convey.

In William Playfair's graph in Fig. 5, the coordinates which have been chosen by the Swiss economist to illustrate the growth of national debt in the United Kingdom in the 18th century produce a thin curve that goes high quite suddenly: the inference that the observer spontaneously makes by looking at this graph is that the growth of public debt has been very fast[15].

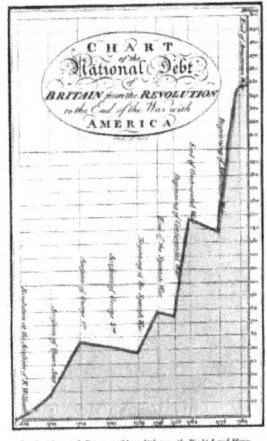

Fig. 4 Playfair's first graph.

---

[15] Playfair (1786). Tufte gives an analysis of this case in Tufte (1983), p. 65.

Nevertheless, Playfair points out that things change if (a) "real" data are inserted in the graph, by modifying when necessary the quantitative data to represent: in this case, in the graph above there was no mention of inflation; (b) another format is chosen: on the x-axis there can still be years, but this time the distance between them is changed; on the y-axis there can still be the debt, but this time the scale chosen will be different. Using this stratagem, the graph assumes a very different shape, as shown in Fig. 6, and the observer does not anymore spontaneously think by looking at the graph that national debt has grown so fast and so bad, though the quantitative information in the graph has been preserved and is even more accurate, and though peaks in the growth are still evident.

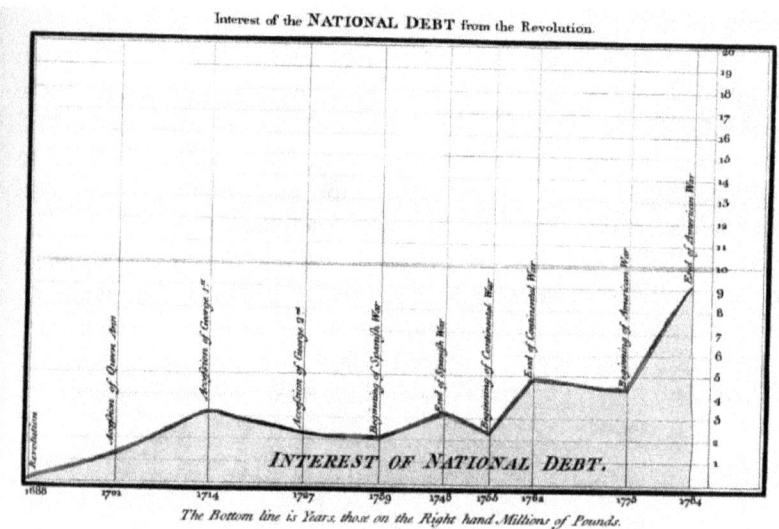

Fig. 5 Playfair's second graph.

A graph therefore activates visual expectations that refer to its behavior. In fact, its global configuration depends on the structure within which the single elements are inserted and become meaningful. The perception of a graph does not demand so much cognitive effort, and these very expectations determine the space of inferences. As a consequence, the information that is extracted from the graph by the observer depends not only from its quantitative content but also from the way in which this quantitative content is organized in a global configuration. The risk is when these visual expectations on the graph promote incorrect inferences.

There are several reasons why for a long time graphs were seen as nothing more than deceitful images. According to Tufte, the first reason was that the people who drew them up were most of the times professional artists and for this reason they were not aware of statistics and often not able to recognize the relevant variables[16]. Secondly, there was a widespread bias towards the salience of statistical data. By contrast, most of all if represented in this format, these data can make explicit relations among variables that would otherwise remain hidden in the linguistic format. Another bias was the idea that graphs were only good for the inexpert reader. On the contrary, interpretation is crucial in graphs even more than in geographical maps.

Therefore, it is not possible to claim that there are graphs that are in principle easier or on the contrary more difficult to understand. Steven Pinker proposes a theory of graphs that relies on two main assumptions: the first is that propositional and structural descriptions of the information contained in a graph are important; the second is that space is "indispensable", since it is the element in a graph that draws together different things such as the user attribution of some predicates, her selective attention and her tendency to create perceptual unities. The extraction of some piece of information is easier for the observer when the visual pattern in the graph by means of which that piece of information is conveyed is easily perceptible, and when the observer is capable of recognizing that it is precisely that pattern that makes the task easier, since it allows her to make more inferences about the information contained in the graph[17].

As much of the empirical literature on graphs shows, the choice of one format instead of another to convey the data depends from the intention of the designer of emphasizing different aspects of these data. A table, for example, is more effective in illustrating the absolute values of a dependent variable, since it would be difficult to make these values explicit in a linear or bar chart. A bar chart can instead be more convenient in illustrating the differences in the values of the dependent variable, since this information corresponds to the height of the bars that can be easily perceived. Finally, a linear graph can be more useful to illustrate trends and interactions, since the observer possesses many predicates to describe the different shapes that a line can assume. Once again, the spatial characteristics of the image constitute the constraints information conveying is subject to, but it is only because the observer knows the conventions used that she can give meaning to these perceptual elements and let them promote new inferences.

---

[16] Tufte (1983)
[17] Pinker (1990)

4. Conclusions

In this article, I have proposed that the use of map implies four components, of which the spatiality of the map is only the first one. In fact, to understand and correctly use a map, it is not only necessary to discern its spatial properties, but also to interpret them in such a way that each location on the map corresponds to a position in the space. Moreover, it is crucial to know that the map can contain us as well.

A map is thus a very interesting object from a cognitive point of view, since it can be used only if other pieces of knowledge that we should already possess are added to it. Furthermore, there is a sense in which a map is always intrinsically incomplete, as also descriptions are, and at the same time it is extremely fascinating because it promotes imagination.

In the second part of the article, I have argued that geographical maps have more recently evolved in other forms of information displaying, such as cortical maps and graphs. These new formats inherit some of the typical features that we have seen for maps but they do also imply new theoretical assumptions and new kinds of interpretations. The universe populated by the different tools that technology now offers to us to display information is now out there only to be explored.

Acknowledgements

I would like to thank Mario Piazza for each of his very useful suggestions. The research was supported by the European Community's Seventh Framework Program([FP7/2007-2013] under a Marie Curie Intra-European Fellowship for Career Development, contract number N 220686 - DBR (Diagram-based Reasoning).

References

Borges, J. L., (1975), A Universal History of Infamy, Penguin Books, London.

Calvino, I. (1972), Le Città Invisibili, Einaudi, Milano.

Dehaene, S., Izard, V., Pica, P., Spelke, E. S. (2006), Core Knowledge of Geometry in an Amazionian Indigene Group, in Science 20, Vol. 311, no. 5759, pp. 381-384.

Gall, F. J. (1819), The Anatomy and Physiology of the Nervous System in General, and of the Brain in Particular.

Harley, J. B. (1992), Rereading the Maps of the Columbian Encounter, in Annals of the Association of American Geographers, Vol. 82, no. 3, The Americas before and after 1492: Current Geographical Research, pp. 522-536.

MacCord, H. (1971), Maps: Poems Toward an Iconography of the West, Kayak Books, Inc., Santa Cruz, Calif.

McCabe, D. P. , Castel, A. D. (2008), Seeing is believing: The effect of brain images on judgments of scientific reasoning", in Cognition, 107, pp. 343-352.

Monmonier, M. (1991), How to Lie with Maps, The University of Chicago Press.

Muehrcke, P. C. and Muehrcke, J. O. (1974), Maps in Literature, in The Geographical Review, LXIV, 3, 317-338.

Parry, J. H. and Keith, R. C. (1984), New Iberian world: A documentary history of the discovery and settlement of Latin America to the early seventeenth century, Vol. 2, New York: Times Books, p. 65.

Pinker, S., (1990), A Theory of Graph Comprehension, in R. Feedle (Ed.), Artificial Intelligence and the future of testing, pp. 73-126.

Playfair, W. (1786), The Commercial and Political Atlas, London.

Racine, E., Bar-Ilan, O. and Illes, J. (2005), fMRI in the public eye, in Nature Reviews Neuroscience, 2005, 6, pp. 159-164.

Shusterman, A., Ah Lee, S., Spelke, E. S. (2008), Young children's spontaneous use of geometry in maps, in Developmental Science, Volume 11, no. 2, p. F1-F7, published on-line.

Stevenson, R. L. (1883) Treasure Island.

Tufte, E. (1983), The Visual Display of Quantitative Information, Cheshire, CT: Graphics Press, 2001.

Notes on Diagrams and Maps

by
Alexander Gerner

Abstract

The paper "Notes on Diagrams and maps" of Alexander Gerner is an exploratory journey that hinges on the very question of what a map is. Gerner draws on the work of Stjernfelt, Krämer and, foremostly, Deleuze and Peirce. It is his contention that these authors' work open up seminal ways of interpreting diagrammatic fixtures common to different notions of the map - as a diagrammatic knowledge tool or an ontological category preceding any representational semiotic operations. The diagrammatic operational and the ontological map reading, though evidently at cross-purposes, should not ultimately be seen as mutually exclusive.

This journey into the map territory in the second part of this paper reflects on contemporary map artists (Beltrán, Artur Bairrio, Thomas Hirschorn & Marcus Steinweg etc.) and their artefacts. Art seems to surge as the place where a map can be accepted as the "map of itself"(Terry Atkinson & John Baldwin 1967). However, that is not to say that this of-itselfness is absent in scientific maps. It is simply not visible, focused on or accepted as such.

The "all knowing map" that Perkins (2008) describes as "scientific", is a seen here as non-sensical. With Deleuze even the "chasm" between different cultures of map use (Perkins 2008) and separated "knowledge regimes" in science and art becomes questionable.

Introduction

This paper is a first exploratory journey that hinges on the very question of what a map is. To guide us, I will draw on the work of Stjernfelt, Krämer and, foremostly, Deleuze and Peirce. It is my preliminary contention that these authors' work open up seminal ways of interpreting diagrammatic fixtures common to different notions of the map - as a diagrammatic knowledge tool or an ontological category preceding any representational semiotic operations.

Several concepts of the map category (Peirce; Stjernfelt 2007/ Perkins 2008/ Deleuze 1995) are proposed for debate. I will be taking a closer look into the notion of a supossed split between diagrammatic spatial-visual maps and operational maps. I take up a suggestion from Deleuze, for whom maps should not be merely seen as visual fixed representations on a plane, as a plant or a scheme to derive knowledge from, but as primary and constitutive, revealing something of their own, "before" or "besides" their "proper" sign activity or function.

Concurrently, I will be analysing definitions of maps that picture them as fundamentally linked to their diagrammatic dimension.

Thus, I will try to sharpen our vision of two rival notions of what a map is. Under the first, which one may call Deleuzian, the map is apprehended as a cartographic entity whose ontological status precedes any given material diagram function. Under the second, the map is grasped as being essentially a diagrammatic tool, bridging the conceptual vs. intuitive gap in knowledge generation. This jorney into the map territory in the second part of this paper reflects on map artists and their artefacts.

Deleuze(1995) locates the map inside a non-representational "plane of immanence", preceding scientific inquiry, but nonetheless a necessary part of a orientational strategy, leading to logical activity in which insignificant or free "playthings" are transformed into "edge tools" for inquiry.

For Peirce, on the contrary, the map is a semiotic subtype of the diagram. As Stjernfelt (2007) and Günzel (2009) also concur, maps share with diagram a common identity, while being able to contain pictorial elements as well.

The tension between these two notions of the map, I claim, should be our starting-point when approaching the phenomenon of the map. Hence, the map is discussed across the diagramatic/ontologic divide, i.e., both

a) as a device, containing several superimposed diagram schemes and pictorial parts for orientation (Peirce)
b) as a cartographic apriori with an ontological claim of its own, preceding concrete diagram schemata (Deleuze) .

The diagramatic and the ontological map reading, I would like to suggest, though evidently at cross-purposes, should not ultimately be seen as mutually exclusive.

This move is crucial and should be borne in mind when considering one of chosen focal points in this paper: the appropriation of maps for artistic purposes, and the hermeneutical regime its "framing " in artistic settings is said to supersede.

The dislocation of maps into an artistic context is usually said to "kill", "switch off" or "deconstruct" supposed conventionalized and pre-existing representational functions of a map, its "conventionalized" status of representation of a "real" object. By framing a map artefact in a gallery, it is generally assumed, "subversive" "and alternative "presentation" is presented. A decoding of an unreflected knowledge regime is alleged to take place. The "all knowing map" that Perkins (2008) describes, immersed in the valley of scientific culture, suddenly vanishes. Deleuze puts all these assumptions into question. The chasm between different cultures of map use (Perkins 2008) and separated "knowledge regimes" in science and art becomes questionable. I tend to hold a similar view. Deleuzes' reflection on a pre-representational nature of maps is partly vindicated by considering maps in artistic settings - and within the trajectory maps of (autistic) children (Deleuze 1995). Maps therein placed show often more explicitly a pre-existing "ontology" by which the concrete artefact was created, while declining to be still considered "optimal representations" or on the epistemic level of meaning construction. With scientific maps, the reverse is rather the case.

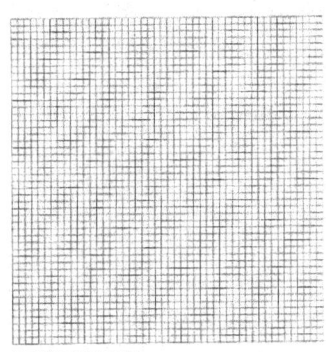

Map 1 Terry Atkinson & John Baldwin "Map of Itself" (1967) Bookprint 14,5x 18 in. from the series "Art& Language" In: Harmon 2009, p.13. The introduction of the grid as a spatial order - " the map of itself"- or the standardized construction of an isotropic, homogenous space in "geometrism"(Bachelard 1994) within the model of "mathesis universalis" reduces and abstracts a real place to an abstract quantifiable, measurable algebraic spatio-temporal extension. The introduction of the grid is shown here as an important device of map-making which map artists explore.

Art seems to surge as the place where a map can be accepted as the "map of itself" (Terry Atkinson & John Baldwin 1967). However, that is not to say that this of-itselfness is absent in scientific maps. It is simply not visible, focused on or accepted as such.

One should consider also the following: scientific map-making is also never free from artistic and formal constraints that may even include general material aesthetic or affective constraints. This constitutes also a good rebuttal to those that defend an out-and-out apartheid between scientific and artistic regimes.

In order to maximize expository value, some formal choices have to be made by the scientific map-makers. An element of artistic preference will also be inescapable at that juncture. These choices may then act as constraints on the step-by-step process of scientific discourse itself.

The introduction of the grid as the standardized construction of an isotropic, homogenous space in modern "geometrism"(Bachelard 1994) within the model of "mathesis universalis" reducing real or qualitative experiential places or topological perspectives (embodied and embedded) to a quantifiable, measurable algebraic spatio-temporal space. This is however merely a passing suggestion – let us not pile more complexity into a subject already fraught with difficulties. It is sufficient for our purpose here merely to state that the artistic rendering of a grid, the legends, the graphical supports, etc, may be of the utmost importance in the actual process of doing science.

In the next section I would like to turn to the diagram category and some important issues linked to the spatial diagram. I will then try to say something about Charles Sanders Peirce concept of operational "diagrammatic thinking" in which the map is seen as a multiplicity of diagram structures, that may include also pictorial parts.

1. On diagrams

With a diagram we may create a new way of relating to uncharted territory. Diagrams succeed when they are able to instil what may be termed an effective dynamic of orientation.

The diagram is a method or a tool used to extend an already existing body of knowledge. The diagram differentiates the initially vague or inchoate in a new way, so that the structural parts of any entity in its rational relations appear and show itself more clear. Diagrammatics is, however, not so much about the concrete shapes and forms of geometrical representation or configuration of knowledge. The diagram also presupposes a specific mode of conceptual reflection. Diagrammatic thinking is concerned with issues and with strategies of transformation of one order towards another. The questions the diagrammatic poses are: what are structures of connectivity and separation? How are such encounters performed? How do they evolve and show forces of change?

A brief foray into art territory brings home this point in a striking manner. In "Personal and Social Order" (2008), Erik Beltrán draws attention to the critical and creative components of map-making in scientific contexts:

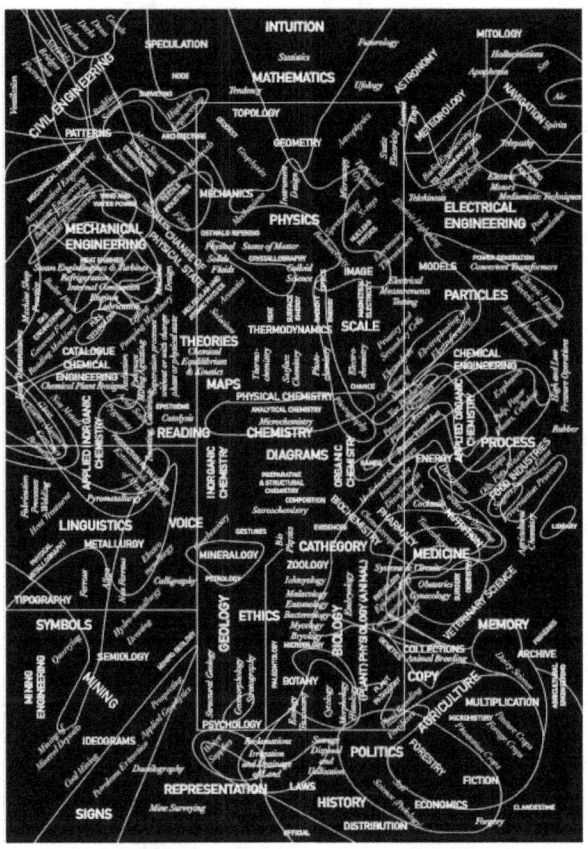

Map 2 Erick Beltrán (2008) "The personal social order"; from: "Calculum Series", Poster b/w Edition of the Barcelona Exhibition at Galeria Joan Pratts December-January 2008/2009 A meta-diagrammatic conceptual playing with orders of thinking, representations and operations in different orders of knowledge, the sub-concepts and lines of which can be related, folded/unfolded in multiple representational and non-representational "clandestine" ways.

In Beltrán's concept map (2008) a non-hierarchical order of a multiplicity of white diagram structures is proposed on a black plane. Rectangular patterns of lines and curved lines partially intersect with conceptual names and its attributed or imagined significations on the black plane. At its very centre- and thus orientating the constellation, the concept word, "diagrams" is readable. Witness that, also present in the same lettering and size , are "ethics", "category", "map" as well as the knowledge disciplines of "physics" and "chemistry , while towards the left the viewer's attention is made to focus on the tension in language between such material techniques of "linguistics" as "calligraphy", "typography", and "voice". The left bottom starts with a ellipse including the concept "signs". On the top of this dynamic there enters a ellipse-like line proposing the concepts "Intuition", "Statistics", "Mathematics. The relational polymath nature of these concept-words are also important when reflecting on how the multiplicity of diagrams can show a structure or a scheme of thought that is perceived and observed as both discursive and iconic .

Map-making, however, is not conceivable without what we bring into it: the blueprint of our habits, the sense of orientation inured into our body (left/right hand etc.). Visual diagrams, as I will try to show in the following chapter are just the end-result of constructed knowledge tools. Their underpinnings – which Deleuze ontologizes (let us bracket for now the question of whether this move towards metaphysics is advisable or not) –lie elsewhere.

1.1 Visual diagrammatics

Mersch (2007) defines diagrams as a proper category (and not a go-between between words and images, or dicursivity and iconicity). According to Mersch, diagrams provide the structure of visual arguments. Diagrams can therefore be defined as visual-graphic schemata, that inform and perform arguments by the medium of the visual. Through diagrams, arguments can be inferred, proofed, refuted and hypothesized (see: Heßler/Mersch 2009, 31). Kramer (2010) invites us to think, in a similar fashion, on an "epistemology of the line", through an extended discussion of the uses and functions of all its different forms (arrows, linking structures, Jordan curves, folds, knots etc.) Kemp (1974) offers an historically- embedded analysis of the place of visual diagrams in the western inquiry tradition by a reflection on the uses of 'disegno". A good way to sum up all these contributions to the study of diagrams, is to say that they all highlight the creative tension between the visible and the knowledgeable. Such tensions are what makes visual diagrams such a worthwhile and rewarding field of study. Both Bredekamp (2005:2007) and Bippus (2009) take this reflection one step further, by turning their attention to the thought-processes implicit in the very act of scribbling a diagram or jotting down a series of dots and lines. As both authors strive to make clear, there are cases in which one may speak with property – in a neo-wittgenstein fashion - of thinking, discovering and creating with the hand. Kleist spoke of the creation of thought by/while speaking. In the same way, Bippus and Bredekamp "close-up" on the relation of eye, hand and thinking (see: Bippus 2009; Bredekamp 2005; 2007) speak of the creation of knowledge through drawing. Diagrams, as I have tried to show in the previous section, are not merely illustrative graphics. Taking up Bippus and Bredekamp probing at face value, one might even say diagrams can actually "show" reflection "caught unaware" – thought in its making, as it were.

Diagrams here do not just include pictorial representations, but can be understood in a broad sense. Krämer (2003) goes so far as to suggest that writing itself is best seen as being already a hybrid construct in which the linguistic and the iconic, telling and showing intersect . She develops this insight in a further refinement with her notion of "Schriftbildlichkeit" (notational iconicity). Here the diagrammatic category is interpreted as making thought "visible" by marks that contain both scriptural and iconic elements.

Finally, we close this survey of the state of the art with a pointer from Waldenfels (2004), who stresses that such a category between the sensible (intuitive) and the intelligible (conceptual) could not exist without the capacity for non-automatic, non-programmable constitutive creatural attention, not bridging the gap between the intuitive and conceptual sphere, but making it explicit to be thought and described.

1.2 Charles Saunders Peirce on diagrams

"Remember it is by icons only that we really reason, and abstract statements are valueless in reasoning except so far they aid us to construct diagrams" (The Logic of Quantity, CP 4.127, 1893)

According to Peirce, diagrams are what he terms eidetic operations or icons of relations. Diagrams for Peirce are Icons, signs that work operationally and therefore enable us to learn something new about the world. Exploring unknown territory, real or imaginary, calls for the drawing of lines, paths and trajectories.

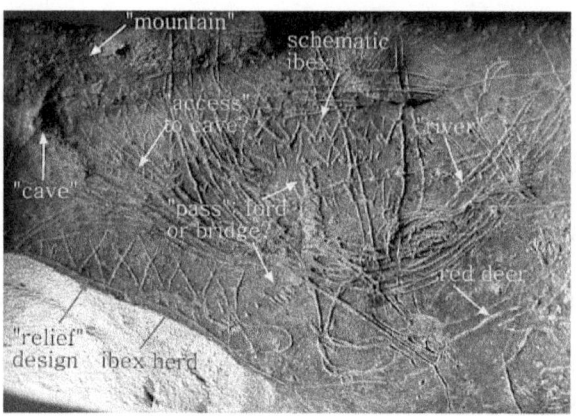

Map 3 Most probably -as we know so far - the oldest material map diagrams were carved 25.000 years ago into mammoth bones and stones as primary human knowledge tools; tools of orientation and memory; directing chaser to places of quarry and hoards of foodstuffs.

However, defining the map not as a representation of a part of the Earth's surface as in modern cartography naturalizes and thus universalizes the map category; "it also obscures its origins in the rise of the state; and it ignores its role in the establishment and maintenance of social relations in those societies where it exists."(Wood; Krygier 2009) It is, however, not our purpose to conflate material map artefacts in all its specific uses in a life-world situation with one specific projection type, or reducing maps to the modern science of geographic mapmaking. On the other hand it is also not my aim to state that by the map only fundamental human abilities of orientation, wayfinding, and other features of spatial intelligence are assumed. In the image above, we can see what could be called a proto-map for orientation in a specific life-world. Recently, the team of the Spanish archaeologist Pilar Utrilla (Utrilla et. all 2009) of the University of Saragossa deciphered what constitutes the oldest prehistoric map-like cartographic engraving in western Europe. This was found in 1993 on a hand-sized stone (1kg) in a cave in Abauntz, Navarra. In the above map-like stone the complex etchings engraved around 13,660 years ago, probably by Magdalenian hunter-gatherers in various times and using various diagramatic design styles, superimpose the discovery's reference point of the „mountain" San Gregorio with animal layers and other geographic layers. For the complementary account of a) mental maps and b) material map(or map-like)-artefacts for orientation in two anthropological models based on ethnographic research to account for the "wayfinding" ability of early humans see: Kirill Istomin and Mark Dwyer (2009).

Peirce also underlines that the observation of diagrams is essential to all reasoning - even if no auxiliary or transformative constructions (manipulations) are performed - there is always a step from the general to a singular statement in deductive reasoning.

Diagrams thus are for Peirce pilots of complex relations, that reveal new knowledge about the world we pilot through by introducing or making explicit hidden or new elements of thought. In peircian epistemology, therefore, diagrams are crucial, since they highlight not only how deductive reasoning is operative, but also can elucidate the very nature of diverse forms of "abduction" (see: Hoffmann 2007).

In Peirce, diagrams are defined as "skeletal icons"(Stjernfelt 2007), representing their object analyzed into parts among which >> rational relations << hold, whether implicit or explicit.

One has, however, to be careful not to adopt a trivial definition of similarity when speaking about diagrams as part of iconicity. Similarity cannot be equated to "identity" with the object itself. Nor can it be psychologised to refer to merely subjective judgments or feelings of resemblance (see: Stjernfelt 2007). For Peirce it is by the icon that knowledge about an object grows:

> *"For a great distinguishing property of the icon is that by the direct observation of it other truths concerning its objects can be discovered than those which suffice to determine its constructions."*(CP 2.279, 1895)

This is what Krämer calls the operationality criteria (Krämer 2009). Thus meaning that icons are signs from which more and new information by observation and manipulation can be derived than what sufficed their construction (Stjernfelt 2000, 2006, 2007). In Peirce's tripartite taxonomy of signs (icon, index, symbol) in the "Syllabus" of 1903 the diagram is treated as a special class of icon .The diagram is defined by similarity to the object that it represents and performs. The diagram presents its object by a "skeletonlike sketch of relations" (Stjernfelt 2007).

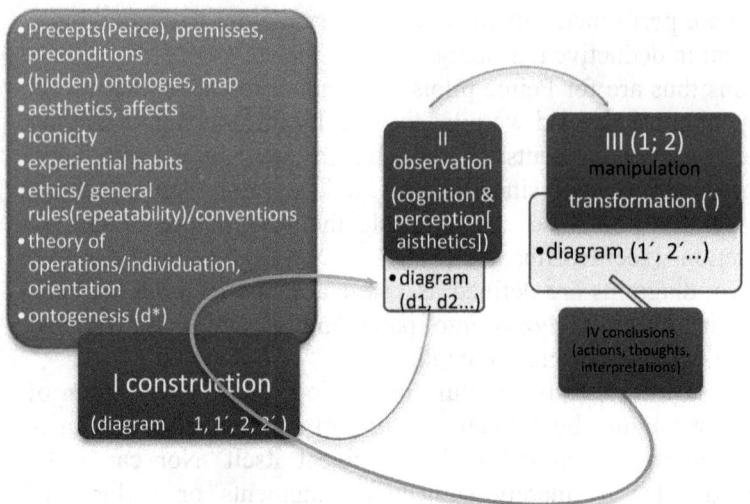

Map 4 (Scheme of the Operational Diagrammatic (blue sections) and the ontological apriori notions (red sections) for diagram operations by A. Gerner inspired by Peirce/Stjernfelt).

For Peirce the operational definition of diagrammatic thinking is the following:

> *"We form in the imagination some sort of diagrammatic, that is, iconic, representation of the facts, as skeletonized as possible. The impression of the present writer is that with ordinary persons this is always a visual image, or mixed visual and muscular (...). If visual, it will either be geometrical, that is, such that familiar spatial relations stand for the relations asserted in the premises, or it will be algebraical, where the relations are expressed by objects which are imagined to be subject to certain rules are, whether conventional or experiential."* (CP 2.778)

> *"By diagrammatic reasoning, I mean reasoning which constructs a diagram according to a precept expressed in general terms, performs experiments upon this diagram, notes their results, assures itself that similar experiments performed upon any diagram constructed according to the same precept would have same results, and expresses this in general terms. This was a discovery of no little importance, showing, as it does, that all knowledge without exception comes from observation."* From Peirce´s application to the Carnegie Institution" (dated July 15, 1902)[From Draft C (90-102)].

Schematically the diagrammatic thinking process is shown above in this meta-diagrammatic map, that would be better exemplified in a animated moving image of thought in which the arrows and the schematic fields of grey (results) and blue (operations) and the precepts and preconditions (red) the immaterial operations(white notations) and the material results(black notations) diagrammatic parts 1,2,1′2′are seen as momentary fixations and directions of a developing and changing process of diagram experimentation.

A map can be seen operationally as a subtype of the extended peircean operative diagram category (Stjernfelt 2007, 105-107; Stjernfelt this volume), in which the diagrammatic principles show up in its use of operations we conduct with the map. Stjernfelt names three of a multiplicity of possible operations that can be experimented with maps as orientational tools: "We may for instance (1) find a route between two localities, (2) determine a distance or an area, (3) recognize landscape forms- on so on"(Stjernfelt 2007, 105) The map-maker and the map manipulator don´t have to be the same entity. Which also obviously means that the intention of the mapmaker and the use, the map-manipulator makes of it, have not to be identical. Stjernfelt reminds us of the experimental part in transformations conducted with maps, "fulfilling the demands for revealing truths not stated in the construction of the diagram" (ibid.) This also shows that the material artefact object in itself is not diagrammatic: only by being used in such-and-such a manner does such a qualification accrue .One has to think of diagrammatics, including eidetic processes, as open-ended.

2. Notes on maps in contemporary art

2.1 Against the fictive scientific "all knowing map"

A 1:1 scale map of the solar system would not just be impossible to make. It would be also senseless. A fully detailed "map" giving "all" information of a world would be impracticable from the point of view of economics of reasoning and impossible to be handled. Perkins' scientific map idealization of "the all-knowing map" (Perkins 2008, 156) is a mere ideal construction of a non-existing reality. A map should be handleable or operational by definition. A map is a helpful economical cognitive tool to orientate and navigate in an ocean of facts that is unfathomable. The context-smashing that occurs by placing a map artefact derived from geography in artistic contexts may help us realize this. By such a dislocation, certain conventions, rules, grammars, logics that are part and parcel of map-making can be made explicit.

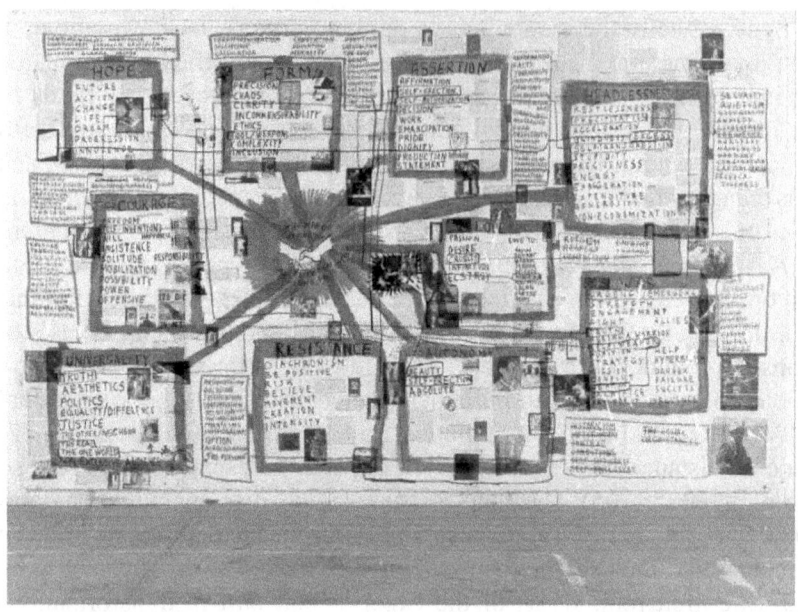

Map 5 Thomas Hirschhorn & Marcus Steinweg: The Map of Friendship between Art and Philosophy (2007), cit. in: http://artnews.org/marcussteinweg/?i=0.

This is a generative "linking machine" and not a fixed definite linkage map, that plays amusedly with our attention, and steers on multiple levels with possible relations of the entities envolved: lines, colours, word-concepts, etc.

An art-piece like the Map of Friendship of Art and Philosophy helps us to realize, by the very deconstruction of diagram schemes it offers, that maps "work "not only by steering our attention into an "intensional" zone. They also de-terretorialize our thoughts from a habitual zone of experimentation towards untrodden territory. And this is precisely what makes them subjects par excellence of contemporary aesthetic discourse. For the philosopher Steinweg the map should not be exlained nor interpreted, but the necessity of the shown concepts should be reflected upon here in the relation of art and philosophy.

While the "context-smashing" of maps in art may provide bridges to Deleuze's insights, thinking of maps in science as "context-feeding" or "context-refining" provides a link to a characteristically peircean diagrammatic reasoning. Thus, reflecting on the place of maps in cultural-artistic and in scientific contexts also brings to light serious differences within the philosophical debate – namely the debate between a position that has not dispelled from his discourses metaphysic undertones - Deleuze-

and a philosophical position that has jettisoned most of its ontological cargo in order to align itself squarely with "scientific" discourse – in the sense of operationality, Peirce.

After saying something about the twin topoi of maps and diagrams in contemporary artists' discourse, in the last part of the paper I will try to pursue some of these default lines in the philosophical debate –focusing again on Deleuze and Peirce.

2.2 Artists as partial mapoclasts: displacements

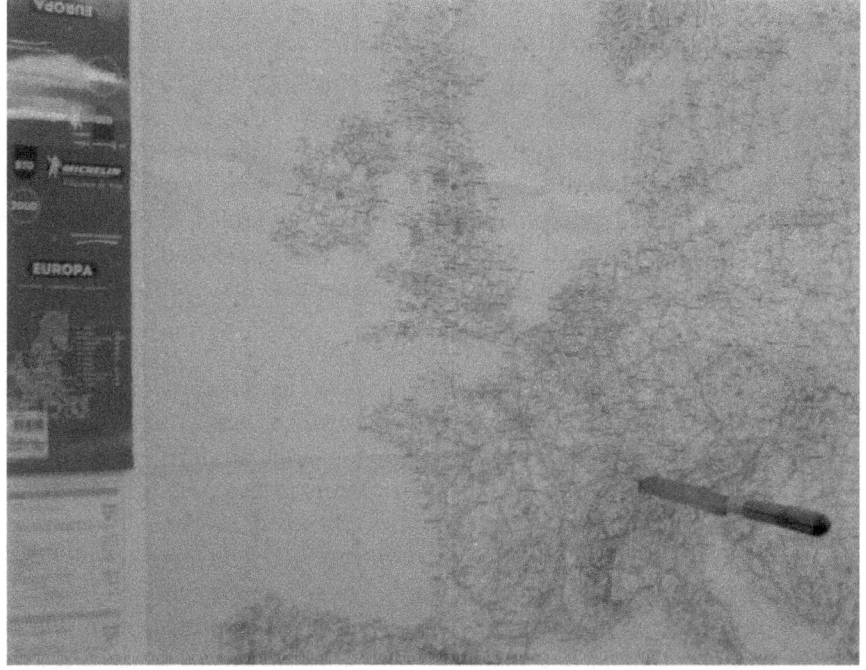

Map 6 Artur Barrio "Uma faca lançada de um ponto qualquer de Portugal sobre um ponto não qualquer/A knive launched from a random point in Portugal to a non- random point", 1975. Europe Map with kitchen knife, 100x154 cm. Collection of the artist.deposited at the Serralves Foundation Porto Photo of the exhibition "Portuguese Artists Abroad" 2008/2009 Lisbon Museum of Electricity
(Photo: A. Gerner 2008).

The localisation that the title (or language diagram) refers to, draws a relation between the visual spatial artefact of this map as a 2-D plan towards the 3-D object of the knife that fixes a reference point in Switzerland on the map plane, cutting up the plane in a brute act of penetrating the 2D plane. The general and abstract orientation possibilities that the map represents is concretized by the individual gesture of thus dramatically fixing an end-point, anchoring the orientation and the visual focus of the observer of this map-artefact. The vagueness of the starting point indexed in the title "a knife launched from a random point in Portugal" is opposed to the point at which the knife cuts the map. One can speculate that the cut or partial damage that the knife provokes is itself a transformative and "context-smashing" act that serves to heighten by contrast the flatness and neutrality of the abstract western "episteme" of mathesis universalis.

Discussing "maps as artistic practices" Perkins (2008) gives several unsystematic examples of map use and map function that "encourage a performative encounter" (Perkins 2008). In Perkins´ view, the map is not just a visualization or a spatio-temporal object. For Perkins this is borne out in the manifold allusions and references to maps in "surrealists, pop artists, situationists, land artists, conceptual artists, community artists, digital media artists and live artists" (Perkins 2008, 156). Such "art-embedded" reflection on maps may include the following artistic treatments and "interventions" on maps:

> *"(...)fragmenting known maps and rearranging them in novel ways; juxtaposing far with near; distorting space into a relative or egocentric form; changing orientation; manipulating projection, scale and generalization to infringe accepted mapping standards; drawing on standard cartographic tropes such as the border, or naming to question social norms; abstracting and over-coding a known form; employing recognisable country shapes in new ways; shifting novel conceptual frames onto familiar icons such as the globe or tube map; and mapping onto different media so as to ask questions about the world or our identities"*(Perkins1 2008, 156).

---

[1] Perkins, C. (2008). "Cultures of Map Use". *The Cartographic journal*, Vol 45, 2, 150-158.

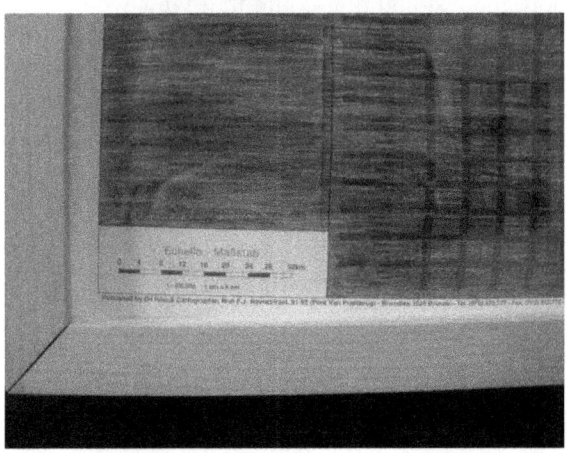

Map 7 Marco Godinho (2007) Série GR# 1-4, Drawing, Pens (BIC black, blue, red, green)
Map Region (Sarre-Lor-Lux-Trier) Palatina-Valnia Collection
of the artist Exhibition view "Portuguese Artists Abroad
Map 8 Detail view: Photo : A. Gerner (2008).

Four equal maps of the same area ,the Grande Région of Luxemburg, France and Germany are successively painted over in a single colour using Bic's famous ball-point pens. The partial re-picturalisation with one single colour assigned to each country, spares out the scale and the Grid A,B, C, (…)1,2, 3,(…) and thus shows again a part of the diagrammatic structure of map-making. One may observe in passing, that a map completely filled with monochromatic colour plane (as the monochromatic overpaintings of the Malevitch) ceases to be a diagrammatic map tool: it is only by contrast and saliency that a map remains a map.

Does all this mean that "art" creates a "new" object? As we have hinted previously, the classic deleuzian move would be to question, not the boundaries between artistic and scientific frames of reference, but their relevance. The philosophical importance of the opposition of artistic praxis to scientific praxis of map-making and map use is fundamentally questioned. Not so much the absolute creation but the displacement of existing planes would be responsible for change. The really hard question is whether these techniques and strategies that artists apply "explicitly" wouldn't also be present "implicitly" in scientific map-making and map-manipulation as well.

A sceptical rejoinder to this is that talk of "implicit" and "explicit" in philosophy is fraught with dangers, for it compromises us sooner or latter with acknowledging "ontologies". Deleuze insists on locating the quid of the map category (Deleuze 1995) before any rational representative order of the diagram. A more modest and non-committal view would be the one that, while admitting that the orthodox/standardized use of maps is of course dependent on the intention of the analysed frame (artistic vs. scientific or expert vs. novice map-user), still acknowledges a commonality above and beyond "cultures of use" in the conceptual and philosophical problems maps raise.

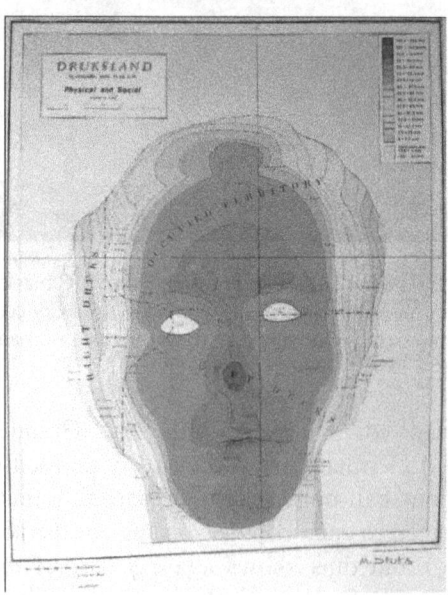

Map 8 Druksland–Physical and Social 15 January 1974, 11.30am 1974
Offset print (unnumbered edition) 18 x 15 inches cit. in:
http://itp.nyu.edu/isco/1/img/IMG_1426.JPG.

The juxtaposition of the personal and political orders in the map of the Israeli artist Michael Druk : different pictorial colour elements, an explanatory legend, a scale of distance, a title and symbolic fixations of topographic areas as for example the "occupied territories". His series of self-portraits metamorphoses geographical into psychographic maps (Debord) and vice versa. The tension in the relation between the topography of the territories in its occupied and unoccupied parts is rendered here as body orientation (left Duks right Duks etc.), and the split between the mediating, rational self and his sub-conscious, irrational part: not a cartesian ego of self-sovereignty, but the map of a "fractured I, of a dissolved ego"(Deleuze 1994,194) shows up. With Deleuze we enter a plane that Peirce leaves ill- reflected. Deleuze's can be seen as a complementary task, a "topo-ontological" not a "logical" one . Deleuze purports to investigate the hidden assumptions, the orientational "map- ontologies", or territories, that diagrams are "constituted" within. The "pure rationality" of a diagram as logic of relations is put into question, and seen as applied in a map of orientation before any representational order or meaning construction. For Deleuze, a kind of cartographic ontology precedes, the diagram category, while for Peirce the map is seen as a concrete subtype of the diagram.

> *"The diagram is no longer an auditory or visual archive but a map, a cartography(...). It is an abstract machine. It is defined by its informal functions and matter and in terms of form makes no distinction between content and expression, a discursive formation and a non-discursive formation.*
>
> *It is a machine that is almost blind and mute, even though it makes others see and speak (...)If there are many diagrammatic functions and even matters, it is because every diagram is a spatio-temporal multiplicity."* Deleuze 1986 1988,34

It is not therefore merely a matter of deciding, whether the map is indexical or related with a material object (the territory) or whether it is the territory itself (see: England 2001). For Deleuze maps are not mere indexes of a res extensa. They are also "maps of intensity":

> *"Maps (...) are superimposed in such a way that each map finds itself modified in the following map, rather than finding its origin in the preceding one: from one map to the next it is not a matter of searching an origin, but of evaluating displacements. Every map is a redistribution of impasses and breakthroughs, of thresholds and*

enclosures, which necessarily go from bottom to top (...) Maps should not be understood only in extention, in relation to space constituted by trajectories. There are also maps of intensity, that are concerned with what fills space, what subtends the trajectories (...)" (Deleuze 1995, 61).

Therefore, the philosophical pregnant point for Deleuze is not that artistic interventions on maps "kill" and deconstruct "identities" and the conventionalized and pre-existing logic of representational relations and symbolic practices, offering "clandestine" counter -proposals to an "all knowing map" (Perkins 2008). For a pragmatist, the natural medium that suggests itself when discussing maps or diagrams is the scientific one.

Deleuze discourse, on the other hand, is concerned with an "ontological" prius that "pragmatist" talk of "roles" and "function" is said to obscure. To talk of "roles" and "functions" is somewhat to forget "essences". Deleuze sees the proper task of philosophy to bring to light a more fundamental order of phenomena- the hidden constitution of such "things" as maps. Herby we have to start to investigate the dynamic aspect of a genetic cartographic a priori of orientation itself (see: Stegmaier 2008; Farinelli 1996, Weigel 2002) in which a posteriori diagram artefacts or perceptible structures of relations are produced in order to transform or displace any kind of orders (visual-spatial, topological, affective etc.).

As Karl Schlögel in his brilliant book "In space we read time" proposes, the necessity to make maps is therefore summed up in the following line:

> *"Always when a world comes to an end and a new is initiated, is the time of the map. Map-times stand for the transformation from one order (of space) towards another."*(Schlögel 2003, 87; brackets A.G.)

Deleuze and Peirce offer respectively an cartographic- ontological or a logical-operational decoding of this.

References

Bachelard, G. (1994). The Poetics of Space. Boston: Beacon.

Bassett, Th. (1998). "Ingenous Mapmaking in Intertropical Africa". David Woodward & Malcom Lewis(Eds.). Cartography in the traditional Africa, American, Arctic, Australian, and Pacific societies. Chicago: University of Chicago Press, 24-48.

Bippus, E. (2009), Skizzen und Gekritzel. Relationen zwischen Denken und Handeln in Kunst und Wissenschaft. Martina Heßler, Dieter Mersch (Eds.) Logik des Bildlichen. Zur Kritik der ikonischen Vernunft. Bielefeld: transcript, 76-93.

Bredekamp, H.(2004). Die Fenster der Monade. Gottfried Wihelm Leibniz' Theater der Natur und Kunst. Berlin: Akademie Verlag.

Bredekamp, H. (2005). Die zeichnende Denkkraft. Überlegungen zur Bildkunst der Naturwissenschaften. Jörg Huber (Ed.). Einbildungen (=Interventionen 14). Zürich, 155-171.

Bredekamp, H. (2007). Galilei der Künstler: der Mond, die Sonne, die Hand. Berlin.

Debord (1955). "Essay: Introduction to a Critique of Urban Geography." Online:http://library.nothingness.org/articles/SI/en/display/2.

Deleuze, G. (1995[1993]). "What children say". Gilles Deleuze, Essays Critical and Clinical. Transl. by Daniel W. Smith and Michael A. Greco. Verso: London, p.61-67.

Deleuze, G. (2006[1988]). The Fold. Leibnitz and the Baroque. Continuum: London.

Deleuze, G. (1988[1986]). Foucault. trans. Sean Hand. Minneapolis: University of Minnesota Press.

Descartes, R. (1992). Meditationes de prima philosophia. Lateinisch deutsch. Hamburg: Meiner.

Dieudonné, J. (1969). Foundations of Modern Analysis. New York: Academic Press.

England, J. (2001). "The Map is not the Territory".Essay by Jane England. Jane England & Co.: London.

Farinelli, F. (1996). "Von der Natur der Moderne: eine Kritik der kartographischen Vernunft". Dagmar Reichert (Ed.). Räumliches Denken. Zürcher Hochschulforum, Vol. 25, Zürich: Vdf, 267-302.

Farinelli, F. (1998). "Did Anaximander Ever Say (or Write) Any Words? The Nature of Cartographical Reason." Philosophy and Geography 1(2), 135-144.

Harmon, K. (2009). The Map as Art. Contemporary Artists explore Cartography. Princeton Architectural Press: New York.

Günzel, S.(2009). "Bildlogik- Phanomenologische Differenzen visueller Medien". Martina Heßler, Dieter Mersch (Eds.) Logik des Bildlichen. Zur Kritik der ikonischen Vernunft. Bielefeld: Transcript, 123-136.

Heßler, M. (2006). "Annäherungen an Wissenschaftsbilder". M. Heßler. (Ed.). Konstruierte Sichtbarkeiten. Wissenschafts- und Technikbilder seit der frühen Neuzeit. Fink, München, 11-37.

Heßler, M.; Mersch, D. (2009). "Bildlogik oder Was heißt visuelles Denken?". Martina Heßler, Dieter Mersch (Eds.). Logik des Bildlichen. Zur Kritik der ikonischen Vernunft. Bielefeld: Transcript, 8- 62.

Hintikka, J. (1983{1980}). "C.S. Peirce´s ′First real discovery′ and its contemporary relevance." E. Freeman (Ed.). The Relevance of Charles Peirce. Chicago: Hegeler Institut, La Salle, 107-118.

Hoffmann, M. (2005). Erkenntnisentwicklung. Klostermann: Frankfurt.

Hoffmann, M. (2004). "How to get it. Diagrammatic Reasoning as a tool of knowledge development and its pragmatic dimension". Foundation of Science 9(3), 285-305.

Hoffmann, M. (2007). "Seeing problems, seeing solutions. Abduction and diagrammatic reasoning in a theory of scientific discovery". O.Pombo, A. Gerner (Eds.). Abduction and the process of scientific discovery. CFCUL, Lisboa, 213-236.

Istomin, K.; Dwyer, M. (2009). "Finding the Way. A Critical Discussion of Anthropological Theories of Human Spatial Orientation with Reference to Reindeer Herders of Northeastern Europe and Western Siberia". Current Anthropology 50, 29-49.

kanarinka (2006a). "Art-Machines, Body-Ovens and Map-Recipes: Entries for a Psychogeographic Dictionary". Denis Wood and John Krygier (Eds.) Art and Mapping: Special Issue of Cartographic Perspectives, 24-30. Online: http://www.nacis.org/documents_upload/cp53winter2006.pdf.

kanarinka (2006b). "Designing for the totally inconceivable: Mods, hacks and other unexpected uses of maps by artists (and other regular people)". Paper presented at the Association of American Geographers Annual Conference.

Kemp, Wolfgang (1974): Disegno. Beiträge zur Geschichte des Begriffs zwischen 1547 und 1607, in: Marburger Jahrbuch für Kunstwissenschaften 19, 218-240.

Krämer, S. (2003). "Writing, Notational Iconicity, Calculus: On Writing as a cultural Technique". Modern Language Notes – German Issue: John Hopkins University Press, Vol. 118 (3), 518-537.

Krämer, S. (2009). "Operative Bildlichkeit. Von der ‚Grammatologie' zu einer ‚Diagrammatologie'? Reflexionen über erkennendes Sehen" M. Hessler, D. Mersch (Eds.). Logik des Bildlichen. Zur Kritik der ikonischen Vernunft. Transcript. Bielefeld, 94-123.

Krämer, S. (2010). " 'Epistemology of the line'. Reflections on the diagrammatical mind". O.Pombo, A. Gerner (Eds.). Studies in Diagrammatology and Diagram Praxis. CFCUL: Lisbon (unpublished).

Lacoste, Y. (1973) "An illustration of geographical warfare." Antipode 5, 1-13.

Mersch, D. (2006). "Visuelle Argumente. Zur Rolle der Bilder in den Naturwissenschaften". S. Maasen, T. Mayerhauser, C. Renggli (Eds.). Bilder als Diskurse – Bilddiskurse. Velbrück Wissenschaft, Weilerswist, 95-116.

Norman, J. (2004). "Review. Mark Graves. The philosophical status of diagrams". Brit. J. Phil. Sci. 55, 801–805.

Schmidt-Burckhardt (2009). "Wissen als Bild. Zur diagrammatischen Kunstgeschichte". M. Hessler, D. Mersch(Eds.), Logik des Bildlichen. Zur Kritik der ikonischen Vernunft. Transcript. Bielefeld, 163-187.

Shimojima, A. (1997). "Operational Constraints in diagrammatic reasoning". G. Allwine, J. Barwise (Eds.). Logical reasoning with Diagrams. Oxford University Press. Oxford, 27-48.

Shin, S. (1994). The Logical Status of Diagrams. Cambridge: Cambridge University Press.

Shin, S. (2003). The Iconic Logic of Peirce's Graphs. Cambridge: MIT Press (Bradford).

Sowa, J. (2001) "Existential Graphs: MS 514 by Charles Sanders Peirce with commentary by John F. Sowa." In: online: <users.bestweb.net/~sowa/peirce/ms514w.htm>, 2001.

Simondon, G. (1995 [1964]). L´individu et sa genèse physico-biologique. Grenoble: Millon.

Short. T.L. (2007). Peirce´s theory of signs. Cambridge University Press: Cambridge.

Sowa, J. (2006). "Peirce's Contributions to the 21st Century". H. Schärfe, P. Hitzler, & P. Øhrstrøm, (Eds.). Conceptual Structures: Inspiration and Application, LNAI 4068, Springer, Berlin, 54-69.

Stegmaier, W. (2008). Philosophie der Orientierung.Berlin/ New York: Walter de Gruyter.

Stjernfelt, Frederik (2000). "Diagrams as Centerpiece of a Peircian Epistemology". Transactions of the Charles S. Peirce Society, 36 (3), 357-384.

Stjernfelt, F. (2006). "Two Iconicity Notions in Peirce's Diagrammatology", in Conceptual Structures: Inspiration and Application. Lecture Notes in Artificial Intelligence 4068, Berlin: Springer Verlag, 70-86.

Stjernfelt, F. (2007). Diagrammatology: An Investigation on the Borderline of Phenomenology, Ontology and Semiotics, Springer.

Stjernfelt, F.(2010). "The extension of the Peircean diagram category Charting the implications of a diagrammatical revolution in semiotics". O. Pombo & A. Gerner(Eds.). Diagrammatology and Diagram praxis. Lisbon (to appear).

Peirce, Ch. S. (1880) "On the algebra of logic," American Journal of Mathematics 3, 15-57.

Peirce, Ch. S. (1885) "On the algebra of logic," American Journal of Mathematics 7, 180-202.

Peirce, Ch. S. (1898) Reasoning and the Logic of Things, The Cambridge Conferences Lectures of 1898, ed. by K. L. Ketner, Harvard University Press, Cambridge, MA, 1992.

Peirce, Ch. S. (1906) Manuscripts on existential graphs. Reprinted in Peirce (CP 4.320-4.410).

Peirce, Ch.S. (CP; references given by volume number and paragraph) Collected Papers of C. S. Peirce, ed. by C. Hartshorne, P. Weiss, & A. Burks, 8 vols., Harvard University Press, Cambridge, MA, 1931-1958.

Peirce, Ch. S. (1976). New Elements of Mathematics [NEM], (ed. C. Eisele) I-IV, The Hague: Mouton.

Pietarinen, A-V. (2006). Signs of Logic. Peircean Themes on the Philosophy of Language, Games, and Communication. Springer, Dordrecht.

Pietarinen, A.-V. (2006b). "Peirce and the logic of image". Online retrived 26.3.2009: http://www.helsinki.fi/~pietarin/publications/Logic%20of%20Image-Pietarinen-Urbino.pdf.

Perkins, C. (2008). "Cultures of Map Use". The Cartographic Journal, Vol 45, 2, 150-158.

Pickles, J. (2004). A History of Spaces. Cartographic Reason, Mapping and the Geo-Coded World. London: Routledge.

Robin, R. (1967). Annotated Catalogue of the papers of Charles S. Peirce. Worcester MA: University of Massachusetts Press.

Toscano, A. (2006). The theatre of production. Philosophy and individuation between Kant and Deleuze. New York: Pelgrave.

Utrilla et. all (2009). "A palaeolithic map from 13,660 calBP: engraved stone blocks from the Late Magdalenian in Abauntz Cave (Navarra, Spain)". Journal of Human Evolution 2009, 57:99-111.

Waldenfels, B. (2004). Phänomenologie der Aufmerksamkeit. Frankfurt: Suhrkamp.

Weigel, S. (2002). " 'Zum ‚topographical turn'. Kartographie, Topographie und Raumkonzepte in den Kulturwissenschaften", KulturPoetik 2 (2002), 151-165.

Wood, D.; Krygier, J. (2009). "Maps" online: http://makingmaps.owu.edu/elsevier_geog_maps.pdf.

"From Images to Diagrams"
Diagrammatic Reasoning in Gilles Deleuze's Film Philosophy and its Relevance for General Media-Theory

by
Christoph Ernst

I. Introduction

While scholars of media theory and media studies tend to agree that film ought to be considered the predominant medium of the 20th century, the question of its potential successors yet remains unanswered. Beyond doubt, digitalization helped pave the way for filmmaking to enter the 21st century. Nevertheless a lot of people remain more or less uncertain about the implications of digitalization for the medium's future development. The major keywords in this ongoing debate are terms such as "hyperrealistic images", "digital imagination," or "film as painting." What seems to be at stake here is one of the most crucial notions that has been at the heart of cinematography ever since its invention: the indexicality of the recorded images. Digital media allow the production of motion pictures which do not necessarily bear reference to a pre-existing reality outside the medium. Instead, the synthetic, i.e. digital images seem to be just as "real" as – or even more real than! – the original. On the one hand, the technique of digitalization can be used to optimize material that already exists. On the other hand, it might be far more interesting to focus on the possibility of producing digital images from scratch. For semiotics the question is: what kind of iconicity is created here?

Artists like Peter Greenaway and philosophers like Vilém Flusser are enthusiastic about the new possibilities of digital filmmaking.[1] Both Greenaway and Flusser claim that digital media can contribute to the liberation of film from the dictatorship of reality. Ironically, their thesis is the exact opposite to one of the most prominent film theories of the last century, namely Siegfried Kracauer's famous Theory of Film: The Redemption of Physical Reality.[2] In fact, the 20th century history of film is above all a demonstration of the attempt to reproduce reality – an attempt governed by the "iconic-indexically complex" of the moving image. The use

---

[1] cf. Flusser, 1998, Greenaway 2007.
[2] cf. Kracauer, 1997.

of digital media in modern filmmaking marks a shift from mimetic representation to synthetic creation, thereby forcing this complex to break apart. As iconicity prevails, the indexicality of a reality outside the medium is no longer considered a relevant category. "Hyper-realistic" imagery offers almost unlimited possibilities to transform or produce all kinds of realistic imagery. In consequence, different forms of "projective" imagery became increasingly popular.

In the wake of digitalization, the medium's self-proclaimed reference to an outside world has been inverted: Films and movies no longer aim at the mimetic reproduction of reality. Instead they now project their very own reality onto the outside (which has formerly been known as "reality"). For Greenaway and Flusser this leads to a dynamic interplay between film and brain, between cinematographic and neurological operations. Thus, what we call "reality" is not "somewhere out there", waiting for us to be discovered, but produced by the transformation of the imaginative potential by means of cinema. More than ever, contemporary filmmaking can be considered a process of iconic conceptualization. Digitalization allows its medium to refer to abstract knowledge about reality (e.g. physical knowledge) and to project this knowledge in form of digital imagery. Infiltrating all forms of visual imagery (TV, film, photography etc.), digital technology effectively undermines ontological differences between fact and fiction while allowing to produce images based on abstract knowledge (such as mathematical oder physical knowledge) about relations and conditions of objects or facts.

How can we theoretically account for this type of iconic reference? Interestingly a prelude to this question has already been provided by French philosopher Gilles Deleuze. In his second book on film, La image-temps, (1985), Deleuze proposes a rather cryptic category: the thinking-image, a subtype of the image-temps ("time-image"). Deleuze suggests this category in order to focus on the question of how film (as a medium) and the human mind interact. With regard to Kant's "Sublime" and Artauds "Theater of Cruelty" Deleuze argues that modernist movies such as Stanley Kubrick's 2001 - A Space Odyssey may overwhelm human consciousness, thereby exposing it to an "outside" which is in fact not external to the medium, but the outside of consciousness itself. This experience can be not only read as a subversion of something like the natural, hermeneutical attitude towards film. For Deleuze, this radical form of inability in thinking itself most notably illustrates the dependency, or more precise: the interdependency of thinking and film (i.e. the medium in general). Following Deleuze, thinking is reliant upon media in order to externalize, formulate, and embody itself.

Due to this model, what becomes important is the process of translation between thinking and film and vice versa.[3]

In my brief essay, I will try to contrast Gilles Deleuze's film philosophy in La image-temps with a reading of key-propositions of Charles S. Peirce's concept of diagrammatic reasoning. In the first part I shall rudimentarily reconstruct the problem of the interconnection between philosophy and film and the basic problem of Deleuze's category of the thinking-image. Secondly, I will try to illustrate Peirce's concept of diagrammatic reasoning. I would like to argue that diagrammatic reasoning is indeed an interesting concept for the philosophy of film, especially when it comes to describe the "projective" references. My hypothesis is that the issue of Deleuze's "thinking images" indicates a possibility to implement Peirce's diagrammatic reasoning into modern film and media theory. I will propose that diagrammatic reasoning is likely to become a key feature of any future theory of filmed imagery, because of its ability to figurize, observe, and refigurize abstract knowledge in a visual form.

II. Deleuze's category of a thinking-image

In his famous film-philosophy, Deleuze argues that aside from the human brain, the medium of film is also capable of thinking for itself. For Deleuze, the medium of film represents a genuine form of organizing chaos. Therefore, its abilities (e.g. the ability to represent motion in the form of holistic movement) can be used not only to reflect on film, but to think creatively with and within the medium of film. The foundation of Deleuze's argument is identical to various theoretical approaches that also refer to the idea of the existence of autonomous, functionally differentiated, super-structures, characterized by the the distinguished feature of operating with meaning.

Theories like systems theory (Niklas Luhmann), deconstruction (Jacques Derrida) or discourse analysis (Jean-François Lyotard, Michel Foucault) share the assumption of a supposed distinction between self- and hetero-reference as basic operations of systems or structures. Relying on the logic of differences (e.g. as laid out in Derridas différance) they perform a first ("blind") distinction between themselves and the outside ("environment").[4] Internal complexity is then created by re-insertion of recurrent observations (as further operations of distinction) within the space

---

[3] Deleuze (1997) dissmisses the term "translation" as a relevant term for his philosophical practice. Nevertheless the process he has in mind can be described as a translation.
[4] cf. Luhmann, 1992.

marked by the initial distinction (an initial with unknown origin that can thus only be "constructed" by the process of reiteration). In the long run, those assumptions evoke the question of how the relation between the differentiated entity and its outside can be stabilized.

Despite the use of a different vocabulary, Deleuze's methodological approach towards film philosophy basically draws on the same ideas by which he then observes and conceptualizes the interaction between philosophy and film. Given the fact that philosophy and film are conceptualized as being autonomous structures of discourse, philosophy is able to "think film" as well as film is able to "think philosophy".

While Deleuze's approach can easily be compared to systems-theory or deconstruction the crucial question of the nature of the supposed interrelation between two apparently separated entities is a bit more difficult to answer. Especially deconstruction and discourse analysis tend to conceptualize the difference as a radical difference with very limited dialogical possibilities. Based on the idea of a rhizomatic structure of knowledge-creation, Deleuze thus proposes a different answer.[5] Philosophy and film, says Deleuze, are not only in principle comparable with each other; they are in fact interconnected. Thereby, the different discourses become an exterior for each other, but at the same time they are enabled to create actual situations in which they rely on the internal possibilities of their complementary discourse in order to perform their own operations. According to this analogy, both the thinking of philosophy and the thinking of film can be imagined as autonomous systems which rely on each other in terms of structure.[6] It is that interdependency which Deleuze calls the image-temps.

By the end of the Second World War, a new type of movie-image emerged. In the course of the following events, it has managed to achieve the status of the dominant type of artistic image in film. The antecedent type, the image-mouvement, is characterized by what Deleuze calls the "lien sensorimoteur" – the sensomotoric link[7]. In this paradigm, filmed images are conceptualized in analogy to the embodied actions and perceptions of an individual protagonist in the film. The protagonist's actions and perceptions are the main criterion for the filmed image. With the image-temps the sensomotoric link disintegrates. As actors are no longer a relevant point of cinematic reference, the movement of the image becomes its own reference. Thereby, the unity of action and perception dissipates, creating new

---

[5] cf. Deleuze/Guattari, 1976, Welsch, 1996, pp. 355-371.
[6] Deleuze by no means proposes an instrumentalistic media theory like e.g. Marshall McLuhan.
[7] cf. Deleuze, 1983, pp. 83-103.

possibilities for reflexive structures within the medium. Movement and time are no longer alignend, providing Film with the opportunity to focus not only on the external movement in the world, but on all kinds of internal and abstract operations such as the mental operation of the reflection of perception, remembrance and oblivion, questions of truth and untruth – and: thinking itself and the relation of concepts and pictures.

In the case of thinking, the main contrast between the image-movement and the image-temps can be illustrated by Deleuze's reading of Sergej Eisensteins "Montage of Attractions."[8] The main focus of Eisensteins montage technique results from the collision of contrasting images that provokes a "third image" within the imagination of the viewer. Especially in his concept of "intellectual montage", Eisenstein considered the third image the product of a dialectical process of reasoning. Contrasting shots are regarded as thesis and antithesis, while the third image in the brain of the viewer is considered the synthesis. Despite the fact that Eisenstein's movies operate within the setting of the image-mouvement, Deleuze welcomes Eisentstein's programmatic editing-style and its ability to realize a philosophical content by a circulation of meaning between the director, the movie, and the viewer. According to Deleuze, the explicitly active role of the viewer in the process of the construction of meaning highlights Eisenstein's approach as the first emergence of a thinking-image. From Eisenstein on, film is no longer considered a secondary medium to human cognition, but a primary medium for human thought; film is a medium by which genuine cognitive potentials are stimulated.

But while Eisenstein's aim was to manipulate the consciousness of the spectator for revolutionary purposes, Deleuze argues that in Eisenstein's montage technique the agitation of the spectator necessarily fails, thus bringing forth a quite different "revolutionary" development:

> *"Le mouvement automatique fait lever en nous un automate spirituel, qui réagit à son tour sur lui. L'automate spirituel ne désigne plus, comme dans la philosophie classique, la possibilité logique ou abstraite de déduire formellement les pensées les unes des autres, mais le circuit dans lequel elles entrent avec l'image-mouvement, la puissance commune de ce qui force à penser et de ce qui pense sous le choc: un noochoc."*[9]

The fusion of thinking and film contains a moment of genuine shock concerning the concept of human thinking: Film offers insight into all

---

[8] cf. Deleuze, 1985, pp. 203-213, Eisenstein, 2006, pp. 9-14.
[9] Deleuze, 1985, pp. 203-204.

aspects of thinking. Not only can it be used as a medium that helps visualize the regular performance of understanding as a faculty of the human mind. Beyond that, it also has a certain "philosophical" ability to visualize phenomenons apart from the primary relation between perception (or intuitions) and understanding. Read as a parallel to the surrealistic concept of "psychic automatism", the third image has to be regarded as a shock because of its epistemological quality. It delivers an insight into the interconnectedness of the human mind with the medium used to create thoughts. The natural status of human thinking is confronted with a medial outside, represented by the medium and its visual and narrative potentials. Consequently, the medium turns out to be a necessary instrument that helps unfold the genuine potential of human cognition.

But cognition's reliance on a medium turns out to be ambivalent. Since thought and medium are complementary domains, they distinguish themselves from one another by means of mutual transcendence. In consequence, every reliance on the medium of film produces a difference to the medium film, thereby undermining the notion of identity between thinking and the medium. As with the interaction between thought and language, there remains a gap between thought and film. Film needs the viewer as much as the viewer needs the film in order to articulate certain ideas. Both systems – film and thought – depend on each other for the simple reason that there is a fundamental difference between them.

For Deleuze, Eisenstein is the first of the great auteurs who adresses this paradox in his aesthetics. The cognitive potentials of the viewer have to be provoked by the filmed image in order to reach a certain rational conclusion. Out of the collision of two antithetic images emerges a critical thought in the viewers mind that accomplishes the dialectical synthesis. Another director, Alfred Hitchcock, made use of the same idea in a much more complex way, namely in his deductive "whodunit"-structures. Hitchcock, no longer trying to limit the idea of a thinking-image to montage, abandoned Eisenstein's dialectical approach. Thus, Deleuze sees Hitchcock's "mental images" as another step away from the image-mouvement into the era of the image-temps. But while a traditional form of the thinking-image lives on in the works of Hitchcock, Deleuze on the other hand claims that it is Antoine Artaud's Theater of Cruelty which has to be regarded as an illustration of a flaw in Eisentsein's aesthetic program. According to Deleuze, Eisenstein's (and Hitchcock's) aesthetics tends to conceptualize the interaction of mind and movie, consciousness and film, as a closed circuit, thus leveling the radical difference between them.

In a reading of Artaud's remarks on the nature of the filmed image, it is the surrealistic program of an écriture automatique that offers for Deleuze an alternative to Eisenstein's dialectical synthesis:

> *"Artaud croit davantage en une adéquation entre le cinéma et l'ecriture automatique, à condition de comprendre que l'ecriture automatique n'est pas du tont une absence de composition, mais un contrôle supérieur unissant la pensée critique et consciente à l'inconscient de la pensée: l'automate spirituel (ce qui est très différent du rêve, unissant une censure ou un refoulement à un insconscient de pulsion)."*[10]

We are being offered two alternatives for an emerging abstract thought in the viewers mind: the dialectical synthesis of Eisenstein on the one hand and the "higher control" (contrôle supérieur) of Artaud on the other – the first being a category of the rational and conscious mind, the second a category of a different kind of ratio emerging from unconscious regions of the mind. Despite his apparent sympathies for Eisenstein, it is Artaud who in Deleuze' eyes managed to capture the crucial problem of the thinking image.

For Deleuze, Eisenstein's aesthetics ultimately lead to a compensation of the difference between thinking and film. For Eisenstein, the difference is a relative difference: mind and film interpenetrate each other. The relation between mind and film is based on the premise of a potential identity between them. Artaud on the other hand regards that difference as a radical difference: mind and film are considered incompatible. In the terms of systems theory: mind and film remain an opaque environment for each other; they operate under the premise of difference between them. Hence, in the era of the image-mouvement film is used to prolong the abilities of thinking, but in the era of the image-temps film represents the mind's outside, confronting thinking not with its inherent potentials but incapacities.

Among many others, perhaps one of the most important questions that can be derived from those ideas is a methodological one: From a film-philosophical standpoint Deleuze's discussion leads to the question of what characterizes a thinking-image in the era of the image-temps. Eisenstein's thinking-images were characterized by a dialectic of concept (mind) and image (film) and thus had a figurative aspect to them. Methodological, the emerging "synthetic" concept, the third image, could be described in terms of rhetoric, especially of metonymy and metaphor. In the era of the image-temps not only the dialectial form of reasoning, but also the figurative

---

[10] Deleuze, 1985, p. 215.

aspects of the thinking-image are being transformed: the thinking-image alters its character from figurative to theorematical:

> *"Tel est le premier aspect du nouveau cinéma : la rupture du lien sensori-moteur (image-action), et plus profondément du lien de l'homme et du monde (grande composition organique). Le deuxième aspect sera le reononcement aux figures, métonymie non moins que métaphore, et plus profondément la dislocation du monologue intérieur comme matière signalétique du cinéma. Par exemple, à propos de la profondeur de champ telle que Renoir et Welles l'instaurent, on a pu remarquer qu'elle ouvrait au cinéma une nouvelle voie, non plus 'figurative' métaphorique ou même métonymique, mais plus exigante, plus contraignante, en quelque sorte théorématique."*[11]

This aspect of Deleuze's argument is a matter of particular interest: Deleuze claims that thinking-images are neither image nor concept and inherently theorematical, meaning that they can no longer be described in terms of (in the widest possible sense of the word) "figurative" images. Applying this theorematical reference, Deleuze describes a new type of reference in which thinking uses a medium to externalize its own potentials by being confronted with its very own incapacities. As for the medium film, this leads to a visualization of its own kind of knowledge in form of moving images. Within the medium of film, knowledge can be observed "in progress", happening on screen and according to the very own, media-specific rules of film. Deleuze exposes this problem mainly as a critical problem for a rationalist understanding of human reasoning. In fact, his critical perspective on traditional ideas about the relation of thinking and film is instructive for film-theory. But besides their critical potential, Deleuze's philosophical considerations hold no solution for the question of how this double-sided epistemological process can be analyzed in a productive manner. Instead we find in Deleuze's text rather provisional considerations about a possible political use of those new types of thinking-images. He simply does not provide us with precise statements about the aesthetics of those images.

Still, his philosophy of film obviously evokes a strong reference to Peirce's diagrammatic reasoning: the idea of theorematical vs. figurative images, then, not only reproduces Peirce's famous division between iconicity in (theorematical) diagrams and (figurative) images. It also implies the ability of the theorematical reference to reduce, reproduce, and transform the

---

[11] cf. Deleuze, 1985, pp. 225-226.

complexity of a potential knowledge of the outside. A question that follows from that assumption might be: Do Peirce's ideas on diagrammatic reasoning help clarifying Deleuze's category? How does film philosophy or media theory benefit from making use of Peirce's ideas?

III. Peirce's concept of diagrammatic reasoning

The use of Peirce's concept of diagrammatic reasoning for the purpose of clarification of Deleuze's argument depends on the questions left out by Deleuze. Furthermore, it also depends on the understanding of Peirce's theory. Given a narrow understanding, one might come to the conculsion that Peirce's diagrammatic reasoning is a theory of what can be done with diagrams. Given a broader understanding, one might come to the exact opposite conclusion. Diagrammatic reasoning might then be understood as a process in which the problem of what can be done with diagrams is only one of many aspects of a general epistemological principle called diagrammatic reasoning. In the latter view, a map is a relevant 'diagrammatic' sign-category as all kinds of schemata are, diagrammatic reasoning can be found in filmed images as well as in mathematical formulae. It doesn't come as a surprise, then, that this wider view is much more fruitful for the discussion of the relevance of Peirce's concept for the purpose of its use with regard to Deleuze's ideas.

In principle, the concept of diagrammatic reasoning is laid out in Peirce's mature sign-theory.[12] Here, Peirce mentions and discusses the concept of diagrammatic reasoning on different occasions, most of them related to the clarification of Peirce's own understanding of pragmatism. One of the many relevant passages is of particular interest. It contains a short synopsis of what Peirce calls 'diagrammatic reasoning':

> *"By diagrammatic reasoning, I mean reasoning which constructs a diagram according to a percept expressed in general terms, performs experiments upon this diagram, notes their results, assures itself that similar experiments performed upon any diagram constructed according to the same percept would have the same results, and expresses this in general terms."*[13]

---

[12] cf. Stjernfelt, 2007, pp. 89-116.
[13] cf. Peirce 1902, Draft C – MS L75.91. This layout of diagrammatic reasoning is obtained from Peirce's so called *Carnegie-Application*. No final version of the text existes, but there are revised and annotated versions.

In the above quotation, Peirce refers to the process of diagrammatic reasoning as he described it in his famous Prolegomena to an Apology for Pragmatizisim.[14] According to Peirce, diagrammatic reasoning is a threefold thought-experiment. (1) Based on a conceptual perception, a diagram of an object is constructed by extrapolation of its main relations. (2) The diagram then becomes the subject of experimental recombination that aims at the evaluation of general rules. (3) Finally, the application of the gained insights into a transformed version of the diagram allows for its comparison with the source-diagram, thereby creating in equal shares new information on the content of the diagram and its related structure.

As we have seen, diagrammatic reasoning is a theory of developing and transforming a diagrammatic structure (as a parametric system of signs). The purpose of diagrammatic reasoning is to gain new information that is not represented in the original diagram. As Peirce himself puts it, its primary purpose is to detect "relations between the parts of diagram other than those which were used in its construction."[15] Diagrammatic reasoning thus describes a deductive process that may result in a creative outcome. Peirce assumes the structures of the diagram to be configured in a way that implies a necessary conclusion which can nevertheless consist of completely new information. As a result of this epistemic operation, diagrammatic reasoning is applied in order to construct out of given premises a diagram, experiment with this construction, and observe the generated information – information that was not contained in the original configuration of the elements and their relations.16 This synoptical definition allows an overall understanding of diagrammatic reasoning, but it still conceals some important details. Those details become visible in the translation of the whole concept in the terminology of Peirce's Semiotics.

In his A Syllabus of Certain Topics of Logic, Peirce offers a well-known definition of the diagram as the second subcategory among three types of signs: hypoicons image, diagram and metaphors:[17]

> *"Hypoicons may roughly be divided according to the mode of Firstness which they partake of. Those which partake the simple qualities, or First Firstness, are images; those which represent the relations – mainly dyadic – of the parts of one thing by analogous relations in their own parts, are diagrams; those which represent the*

---

[14] For lack of space the short version is used here. A more detailed description of the process can be found in Peirce, 1976, IV, pp. 316-319.
[15] cf. Peirce, 1976, III, p. 749.
[16] cf. Hoffmann, 2005, pp. 123-142, Stjernfelt, 2007, pp. 103-105.
[17] cf. Peirce, 1998, pp. 258-299, esp. pp. 273-274.

> *representative character of a representamen by representing a parallelism in something else, are metaphors.*"[18]

According to this definition, the diagram consists of a skeleton-like image of the main relations of an object. The structure of its relations represents a selection of the relations of the object. This similarity in relations defines the diagram not only as an icon of the object, but as an icon to be a sign-class with deductive features: the diagram is an abstraction showing regularities and allowing those regularities to be used in a process of reasoning. As a consequence, the translation of the synopsis of the concept of diagrammatic reasoning in semiotic terms can be read as follows: Peirce's "percept expressed in general terms" has to be considered a symbol, to which the diagram builds a relation, thereby positioning the diagram in the prototypical position of a "type". The general rule given in the prototypical structures of diagram now stimulates a process of "experimentation", comparable to the creation of numerous "tokens". The rules represented in the structures of the diagram (and all their features like rationality etc.) make up a general aspect of the object, thet – because of its generality – obtains a symbolic character.

Hence, the diagram consists "of two parts: a diagram token and a set of reading rules for the understanding of it as a type."[19] The diagram has to be seen as a diagram of general principles (empirically manifest in the common use of diagrams for illustrations of abstract principles such as the planetary movement etc.). Thus the diagram is a construction of a deductive structure as well as it is characterized as a hypothetical notion. In consequence, the reference established by a diagram is related not only to actuality, but to potentiality: the diagram is a hypothesis on the possible status of the constitutive rules of its object. Therefore, it is moved into the space of imaginary possibilities (and by that is the subject of a thought experiment). The process of experimentation recombines the content manifest in the diagrammatic structures according to the rules by which those structures are generated. The observation of that transformation, displacement etc. can provide insight into the configuration of the elements as well as the validity of their defining rules. In other words: related to the original object the diagram creates an quasi-objectivity for epistemic purposes, i.e. a way to think about practical consequences of real possibilities.

---

[18] Peirce, 1998, p. 274.
[19] Stjernfelt, 2007, 97.

As a "vehicle for mental experiment and manipulation"[20] the diagram – according to Peirce – results in a revised and transformed version of the diagram expressed "in general terms".[21] The conclusion of the diagram thus illustrates two things at once: a revised version of the meaning of the object and a revised version of the diagram. Both require the double-sided conclusion to be inherent in the premises of the original diagrammatic outline, thereby allowing a conclusion about the status of the original object and creating new information about the object: "Thus, the steps in diagrammatic reasoning lead from an initial symbol though three consecutive phases of diagrams and to a final symbol."[22] In other words: Peirce's concept of 'diagrammatic reasoning' is a theorematical concept.[23]

Usually, the term "theorematical" covers three different meanings: in psychoanalysis the theorematical content of a dream is perceived differently from an allegorical content because it is undisguised; in the tradition of rhetoric a theorematical form of speaking is an abstract form of speaking that remains different from a figurative form; and in mathematics a theorem is separated from a problem due to the fact that a theorem is itself a consistent set of theoretically well-founded axioms and premises, while a problem yet remains to be clarified.

Apparently, the concept of diagrammatic reasoning includes theorematical aspects form the mathematical as well as from the rhetorical tradition. While the mathematical tradition is present in the deductive nature of diagrammatic reasoning, the process of establishing diagrammatic structures and experimenting with the possibilities for recombination, the rhetorical tradition lives on in the evaluative type of similarity offered by diagrammatic structures. The major epistemological function of the three steps of diagrammatic reasoning – the construction of a diagram, the experimentation with its structures, and the observation of the results within a new projection of the diagram – is a theorematical one.

In principle, the use of Peirce's concept is obvious. It enables us to explain the pragmatic use of diagrammatic signs in general. It also provides us with an idea of the process of learning with the help of signs, or vice versa: an insight into the strong relation between thinking and signs, and into the importance of signs in the epistemic process. Furthermore, the theory holds an interesting answer to one of the crucial questions of epistemology:

---

[20] Stjernfelt, 2007, 99.
[21] See also Peirce, 1976, IV, p. 238: "A *diagram* is an *icon* or schematic image embodying the meaning of a general predicate; and from the observation of this *icon* we are supposed to construct a general predicate."
[22] Stjernfelt, 2007, 103.
[23] Stjernfelt, 2007, 104.

How is it possible to create new information as a type of information that is by definition more than the sum of the elements and their relations (structures) within a given configuration? Critical as this question already is, it also highlights the connections between the concept of diagrammatic reasoning and other philosophical questions both outside of Peirce's philosophy as well as within his semiotics.

Peirce's concept deals with issues of Immanuel Kants epistemology, namely the problem of "transcentental schemata" in the chapter on "schematism" in Kant's Kritik der reinen Vernunft.[24] Within Peirce's own philosophy that problem is connected to the epistemological operation of abduction and the logic of the pragmatic maxim. Like diagrammatic reasoning (and in a way like the problem discussed in Kants "schemata"), all those concepts are developed in order to provide a theory of the concretization of meaning.[25]

As Peirce has pointed out, the process of diagrammatic reasoning is not limited to the human mind but can take place in any semiotic quasi-mind. According to the premises of Peirce's semiotics, the design of diagrammatic reasoning is of holistic nature. One of the many implications of this design concerns the interaction between internal, i.e. mental, signs and external, i.e. material signs. Given a semiotic continuum between internal-mental and external-material signs, the theory of diagrammatic reasoning – in order to be a theory of the concretization of meaning – has to be considered a theory that has been explicitly designed to describe a process of reasoning which bridges the gap between internal and external signs.[26]

In order to develop a perspective for a possible application of the concept of 'diagrammatic reasoning' on Deleuze's ideas, it is important to identify the problems for which the concept provides an answer. His own description of the status of the thinking-image as a theorematical type of image suggests that there might be a link to Peirce's concept of diagrammatic reasoning.

IV. The 'thinking-image' – regarded as 'diagrammatic reasoning'

Deleuze's film-philosophy shifts the focus from the axiom of a relation of identity between consciousness and film to the axiom of a fundamental yet productive difference. Hence, film has to be conceptualized as a medium that not only activates thinking (like Eisenstein's rhetorical

---

[24] cf. Kant, 2002, pp. 213-222.
[25] cf. Pape, 2002.
[26] cf. Shin/Lemon, 2003.

aesthetics suggest) but can be used for specific epistemic purposes, in which, tautological as it might seem, the thought of thinking has yet to be developed within the medium itself. This premise follows from the axiomatic change that Deleuze proposes: film both operates as a medium for the creative production of thought and (by that) as a medium that allows for the subversion of thought. It is thus necessary to combine the two perspectives on the interplay of thought and film: the perspective of a relative and the perspective of a radical difference.

Deleuze seeks that interplay in essayistic assumptions about a deeper form of connectedness between thinking and film ("membrane"). This is where Peirce and the concept of diagrammatic reasoning can be applied, since it can be used to describe the process of displaying the relational structures of content (such as plot structures) and form (such as frame, shot, and editing). But that's not what is useful for a discussion of Deleuzes argument. His notion of a theorematical nature of the thinking-image aims at a mode of reference which is not reproductive and representative (like the image in Peirce), but productive and imaginative (like the diagram in Peirce). What Deleuze seems to be advocating is a diagrammatic use of filmed imagery as a method of thinking with a medium.

Deleuze's thoughts on the thinking-image extensively deal with the problem of the relation between internal and external aspects of thinking. It is one of his concerns to show that this complex relation is based not on identity but on difference. Besides the rather crude adaptation of Peirce's Semiotics in Deleuze's film philosophy, and besides Deleuze's own category of the diagram,[27] one can bring forward another argument, namely that Deleuze's category of the thinking-image can be clarified if being considered a process of diagrammatic reasoning. In order to further illustrate the potential that Peirce's concept of diagrammatic reasoning holds for Deleuze's thinking-image, it might be useful to discuss the issue alongside the three main aspects derived from Peirce's description of the process of diagrammatic reasoning.

Those aspects are the three key-components of diagrammatic reasoning, subdivided here for the purpose of further specification:(1) The construction of diagrammatic structures – a reasoning process constructing a diagram and its own parametric system of coded description; (2) The experimentation with diagrammatic inference – a reasoning process which relies on exemplification as its primary form of reference; (3) The observation of diagrammatic reference – the combinatory work with those structures and their reference in a thought-experiment creating new information.

---

[27] As developed e.g. in Deleuze (1986).

The first criteria, the construction of diagrammatic structures, is related to the problem of rules. To speak of a thinking-image in the era of the image-temps means to speak of an image whose aesthetics is not longer a simulation of a human form of thinking. Instead, any evaluation of a given content under a set of rules can be considered a way of thinking. So the thinking-image is a regulative kind of image, comparable to the semiotic category of a type. The difference between thought and film is no longer regarded as a difference concerning the relation between the image on the screen and the cognition of a human brain in a movie theater. On the contrary, it can be described as a relation between an image-like content and a thought-like rule in the image on the screen, representing a process of concretization of meaning and by that (!) clearing the process of the concretization of meaning.[28] The circulation of thought and film is a circulation between an image-like content and a thought-like rule on the screen. Therefore, it is a diagrammatic image with a paradoxical status: on the one hand it is a diagram of the relation between thought and film (resp. the image and the consciousness of the viewer) on the screen; on the other, it is also the first image to explore this relation and by that at once creating and representing what is happening on the screen. Regarded as a diagram, the thinking-image is the image of a temporal paradox: through a diagrammatic form of similarity it shows an established relation that is created in the performance of showing it.[29] It is the paradox of a constructed iconic similarity: it reveals iconicity as created that is created at the very same moment.

Because of its diagrammatic nature, the thinking-image is characterized as an image form the era of the image-temps. This leads to the second criteria: experimentation by diagrammatic inference. The construction of diagrammatic structures in filmed images is not fully covered if a micro-analytical perspective only deals with the use of obvious diagrammatic structures, like for example by using maps. It is far more profitable to begin an analysis on the macro-level. The fact that the thinking-image shapes a regulative form of aesthetics is accompanied by the issue of its relational character: a thinking-image consists not of a single type of picture, whose characteristics can be described phenomenologically and analyzed hermeneutically. Instead, it is realized in a performative way by relating different poles within the field of possible combinations of the

---

[28] This has an interesting but difficult consequence for the exegesis of Deleuze's theory: it appears this film-philosophy is far more pragmatic in nature than its structuralist origins indicate at a first glance.

[29] As a temporal issue this paradox illustrates not the abilities of diagrammatic reasoning, but of Deleuze's film-philosophy for the application of the concept diagrammatic reasoning.

relation between thought and image. This process takes place as a process of evaluation of other possibilities within the image. By using thinking-images, the medium of film not only enacts a process of thinking but also evaluates the validity and the possibilities of the medium's description of the process (and all of its own possibilities). The issue of thinking within the limited possibilities provided by this specific medium is always associated with the thinking-image. Counterintuitively, this does not lead to the question of whether or not the filmed process of thinking can be an adequate description of thought (which would be an understanding of iconic similarity based on the premises of identity). Instead, it refers to the pragmatic process of an evaluation of which thought can be realized by that specific medium and its inherent rules of presentation (which is an new understanding of iconic similarity based on the premise of difference).

Given the dynamics of the reflexive process of the thinking-image it establishes a self-referential epistemological scene that is related to the third criteria, the observation of diagrammatic reference. The thinking-image then establishs a new form of reference, changing its vector from an assertion of a denotative understanding of similarity to an evaluation of the possibilities of the construction of that similarity. It is easy to see that the main premises of Deleuze's argument are not fully covered by a critique of a representative or ontological perspective and its replacement through a functional perspective. Beyond that, Deleuze faces the necessity to account for two different types of reference. His argument may reveal some of the problems that arise from a naive understanding of a certain form of iconic similarity based on denotation, but Deleuze more or less fails to develop a new perspective for a form of iconic similarity based on exemplification.[30] The thinking-image does not show thinking by representing it as a given denotatum. It is an exemplification because the denotatum of the image is a display of the exact properties of its denotation. Being that construction, it is necessarily a diagrammatic selection, the display of a layout of the possibilities to answer the question of what can be considered as thinking within the medium of film. The thinking-image constructs a form of quasi-objectivity, thereby crossing the difference between thought and image in order to overcome and by that prolong the difference between those entities.

Synopsis and further issues

Let me briefly point out the results of my preceding arguments. My goal was to illustrate the relevance of Peirce's semio-pragmatic concept of

---

[30] cf. Goodman, 1976, pp. 55-57.

diagrammatic reasoning for the further clarification of the category of the thinking-image in Deleuze's film-philosophy. Deleuze's category of the thinking-image is a critique of theoretical assumptions based on a naive understanding of the relation between thought and image. As Deleuze argues, the medium of film is in itself a theoretical entity. On the one hand, the human mind makes use of film as a medium that helps clarify the process of shaping its own thoughts; on the other hand, the medium provides the possibility to create that clarification and thus determines the process. In consequence, the relation between thought and image can only be regarded as a relation of difference, a relation that refuses assumptions about the identity between mind and film. Peirce's concept of diagrammatic reasoning can be brought into play the moment this approach has to describe how the process of constructing thought is realized. It gives a clear account of how this reflexive process takes shape in film. Furthermore, it helps to describe the possible outcome of that process by providing a solid theory of the creative evaluation of information.

The concept of diagrammatic reasoning is a good example for the benefit that modern film-philosophy can obtain by referring to a pragmatic reading of Peirce's semiotics. A comparison of Deleuze's concept of the thinking-image and Peirce's concept of diagrammatic reasoning reveals some striking similarities between them. It is very likely that Peirce's diagrammatic reasoning will play a major role in all kinds of further film-philosophical discussion about the relation between thought and image, film and consciousness. A further analysis and deeper discussion of Deleuze's adaptation of Peirce's semiotics in his film-philosophy might even show that diagrammatic reasoning is exactly the process which Deleuze himself had in mind when he conceptualized his category of a thinking-image. Furthermore, this exemplary essay suggest further applications of Peirce's model of diagrammatic reasoning in media-theory.

Current media-theory relies heavily on concepts and ideas developed in (de)constructive approaches and theories. Just like in Deleuze's film-philosophy, the application of Peirce's equally semiotic and pragmatic philosophy in the context of those concepts can create interesting theoretical insights for central issues of media-theoretical discussions.[31] In this respect, one of the most crucial issues might be the combination of analogue and digital imagery in film. Digital imagery creates an new kind of iconicity, lacking the kind of similarity associated with an ontological or denotative type of reference. Deleuze mentions the issue of digital imagery briefly, although he does not provide further discussion:

---

[31] cf. e.g. Jahraus (Ed.), 2001.

"La figure moderne de l'automate est le corrélat d'un automatisme électronique. L'image électronique, c'est-à-dire l'image téle ou vidéo, l'image numérique naissante, devait ou bien transformer le cinéma, ou bien remplacer, en marquer la mort. Nous ne prétendons pas faire une analyse des novelles images, qui dépasserait notre projet, mauis seulement marquer certains effets dont le rapport avec l'image cinématographique reste à déterminer. Les nouvelles images n'ont plus d'extériorité (hors-champ), pas plus qu'elles ne s'intériosent dans un tout: elles ont plutôt un endroit et un envers, reversibles et non-superposables, comme un pouvoir de se retourner sur elles-mêmes. Elles sont l'objet d'une réorganisation perpétuelle où une nouvelle image peu naître de n'importe quel point de l'image précédente".[32]

According to Deleuze, the main feature of digital ("numerique" resp. "èlectronique") imagery is its ability to become the subject of a process of permanent recombination and variation. Bearing in mind the three main features of diagrammatic reasoning and their relevance for Deleuze's thinking-images, Peirce's diagrammatic reasoning can be considered a valuable concept for the description of the construction of a projective kind of reference and its pragmatic use for the recombinating and evaluating of a given content. In fact, the process of reasoning associated with the thinking-image must be regarded as being closely related to the problems of iconicity in digital imagery. Concerning this matter, two problems are of peculiar interest. (a) The problem of a temporal paradox in digital iconicity: this kind of iconicity creates similarity to an object before the object even exists. (b) The problem of experimentation and evaluation of other possible forms of representation. Exactly because of its hyper-realistic appearance, every digital image can be considered as an image that is based on an assumption of what its object might look like. Digital Realism is a calculation of what reality could look like.

It is quite likely that the use of diagrammatic reasoning in modern film-philosophy is not only relevant for the discussion of the relation between thought and image, consciousness and film. Furthermore, it has to be read with regard to the overall pragmatics of the use of digital imagery in modern media-culture, thereby becoming an issue of great interest for media-theory in general.

---

[32] Deleuze, 1985, p. 346-347.

References

Deleuze, Gilles (1997), Le bergsonisme, Paris, Univ. de France.

Deleuze, Gilles (1983), L'image-mouvement. Cinema 1, Paris, Les Éditions de Minuit.

Deleuze, Gilles (1985), L'image-temps. Cinema 2, Paris, Les Éditions de Minuit.

Deleuze, Gilles (1986), Foucault, Paris, Les éditions de Minuit.

Deleuze, Gilles/Guattari, Felix (1976): Rhizome, Paris. Les Éditions de Minuit.

Eisenstein, Sergej (2006), Jenseits der Einstellung. Schriften zur Filmtheorie, ed. by Felix.

Flusser, Vilém (1998): Vom Subjekt zum Projekt. Menschwerdung, Frankfurt/M.: Fischer.

Greenaway, Peter (2007), "Das Kino neu erfinden", in: Kloock, Daniela (Ed.), Zukunft Kino. The End of the Reel World, Marburg, Schüren, pp. 275-286.

Goodman, Nelson (1976), Languages of art. An approach to a theory of symbols. Indianapolis, Hackett Publishing Company.

Jahraus, Oliver (Ed.) (2001), Bewusstsein, Kommunikation, Zeichen. Wechselwirkungen zwischen Luhmannscher Systemtheorie und Peircescher Zeichentheorie, Tübingen, Niemeyer.

Hoffmann, Michael (2005), Erkenntnisentwicklung. Ein semiotisch-pragmatischer Ansatz, Frankfurt/M., Klostermann.

Kant, Immanuel (2002): Kritik der reinen Vernunft, Stuttgart, Reclam.

Kracauer, Siegfried (1997): Theory of film. The redemption of physical reality. Princeton. Princeton Univ. Press.

Lenz, Helmut H. Diederichs, Frankfurt a. M., Suhrkamp.

Luhmann, Niklas (1993), "Deconstruction as second-order observing", in, New Literary History, 24, 4, pp. 763-782.

Pape, Helmut (2002), Der dramatische Reichtum der konkreten Welt. Der Ursprung des Pragmatismus im Denken von Charles S. Peirce und William James, Weilerswist, Velbrück Wissenschaft.

Peirce, Charles S. (1902), "MS L75. Logic, regarded as semeiotic (The Carnegie application of 1902)",ed. by Joseph Ransdell, http://www.cspeirce.com/menu/library/bycsp/l75/l75.htm.

Peirce, Charles S. (1976): New elements of mathematics, Vol. I-IV, ed. by Carolyn Eisele, The Hague, Mouton, Humanities Press.

Peirce, Charles S. (1998): The essential Peirce. Selected philosophical writings, Vol. 2, 1893-1913, Bloomington, Indiana Univ. Press.

Shin, Sun-Yoo/Lemon, Oliver (2003), "Diagrams", in: Stanford Encyclopedia of Philosophy, http://plato.stanford.edu/entries/diagrams.

Stjernfelt, Frederick (2007), Diagrammatology. An investigation on the borderlines of phenomenology, ontology, and semiotics, Dordrecht, Springer.

Welsch, Wolfgang (1996), Vernunft. Die zeitgenössische Vernunftkritik und das Konzept der transversalen Vernunft, Frankfurt a. M., Suhrkamp.

Deleuze's Rhizome-Thought

by
Catarina Pombo Nabais

> *"Write to the nth power, the n-1 power, write with slogans: Make rhizomes, not roots, never plant! Don't sow, grow offshoots! Don't be one or multiple, be multiplicities! Run lines, never plot a point! Speed turns the point into a line! Be quick, even when standing still! Line of chance, line of hips, line of flight. Don't bring out the General in you! Don't have just ideas, have just one idea (Godard). Have short-term ideas. Make maps, not photos or drawings. Be the Pink Panther and your loves will be like the wasp and the orchid, the cat and the baboon. As they say about old man river: He don't plant'tatos/ Don't plant cotton/ Them that plants them is soon forgotten/ But old man river he just keeps rollin'along".*[1]

How is the world? How does it function? We can think it as the trees: one, two, three, four, eight, always in a dichotomic way, in a binary logic, within the idea of totality. But Nature does not act this way. On the contrary, Nature is a question of grasses, flights, molecules. It is a very different operation. By transbordance, by intersection, by symbiosis, Nature constantly escapes from an organization according to the image-tree. Grass grows and fills empty spaces, uncultivated fields, intervals between stones. The same for thought, says Deleuze. Between well defined forms, between subjects, there is always a populated space where bad grass never tears off completely: invisible population, intangible materiality, particles in relationship of movement and rest, speed and slowness, intensities. Subjectivity is relation, population, multiplicity. Thought is more a matter of grass rather than of trees. It is a composition of multiplicity within the model of the rhizome which is non-hierarchical, heterogeneous, multiplicitous and acentered.

The Oxford English Dictionary defines rhizome as "a prostrate or subterranean root-like stem emitting roots and usually producing leaves at its apex; a rootstock."[2] According to the OED, the word rhizome derives from the Greek (from «rhizoma», mass of roots, from «rhizoun», to cause to take root, and from «rhiza», root), but only dates from the middle of the 19th century. So, rhizome comes from botany, and the most common rhizomes are pieces of ginger.

---

[1] Deleuze, G./Guattari, F., *A Thousand Plateaus* (TP), tr. Brian Massumi, London/New York, Continuum, 2004: 27.
[2] "Rhizome" *Oxford English Dictionary*. ed. J.A. Simpson and E.S.C. Weiner. 2nd ed., Oxford: Clarendon Press, 1989, OED Online Oxford University Press.

The concept of rhizome is central to understand Deleuze and Guattari's point of view over both the world and the thought. In their book titled Rhizome: Introduction (1976) which was lately included as "introduction" in A Thousand Plateaus (1980), Gilles Deleuze and Félix Guattari, together developed an ontology based on the rhizome. There, they expanded the concept of rhizome to all nature, saying that "even some animals are [rhizomes], in their pack form. Rats are rhizomes. Burrows are too, in all their functions of shelter, supply, movement, evasion, and breakout. The rhizome itself assumes very diverse forms, from ramified surface extension in all directions to concretion into bulbs and tubers."[3]

Much about Deleuze and Guattari's rhizome can be understood through its opposition to the image of the tree. In the tree, all branches refer back to a central root in a necessary or essential way. It is a matter of vertical hierarchy. There is always a central point in the tree, a single trunk, from where all the branches grow. This central point is the higher unity of all the branches, it is the Subject that all the branches represent, imitate and reproduce. So, arborescence is a matter of filiation. It is a binary system, based on the dualistic relation of One-Two, i.e. it is a system where the One/Subject-unity represents the model and gives the information to its Two/Object-segments so that they can reproduce and grow. Their filiation implies the verb "to be" (which expresses a centralized entity from where all subjectivity grows), and the exclusive conjunction "either... or" (which means the binary and dualistic relation between that central entity and its representations).

On the contrary, a rhizome is an underground and horizontal root system, a subterranean "canal" that attaches itself to other root systems and scatters in all directions. Precisely because there are no hierarchical relations, the rhizome implies multiple heterogeneous dimensions which refer to each other by alliances. Instead of the verb "to be" and the exclusive conjunction "either... or", the rhizome expresses a becoming which is always a process of the inclusive conjunction "and..., and..., and...". The rhizome "is composed not of units but of dimensions, or rather directions in motion. It has neither beginning nor end, but always a middle (milieu) from which it grows and which it overspills. It constitutes linear multiplicities with n dimensions having neither subject nor object, which can be laid out on a plane of consistency, and from which the One is always subtracted (n – 1)".[4] Instead of the binary logic of the tree, the rhizome functions with the multiple and acentered system of the n – 1, i.e a system where the subject and the object (as the One) are subtracted. Without a central trunk, the

---

[3] TP: 7.
[4] TP: 23.

rhizome connects every point to each other. It exists, not in one extremity from where everything grows, but in the middle, with no centre, with no growing point but only lines of connection, many different points and dimensions, all of which connect to each other in various ways and directions.

But for Deleuze and Guattari, rhizome is an open system of both Nature and Thought. First, instead of an arborescent model which describes hierarchical systems of thought, the "rhizome" (or lateral, multi-forked root system) suggests the nomadic movement of thought, of a thought in process. Secondly, instead of a static model, the rhizome implies the concept of "becoming" or "performance", which means a dynamic and heterogeneous way of thinking. Ideas are dynamic events or "lines of flight" which take us into an endlessly bifurcating system.

Now, does this give place to the first dualism in a thought which aims at exceeding any duality? Yes, it does. However, the tree's operation is not exhausted in a binary logic. Let us look at its branches and its roots: there is something of rhizomatic. Also, in the rhizome, there are tree structures. The rhizome is the alternative to the dualistic and binary logic, not as its opposition (which would mean a dualistic system), but rather as a new type of logic: the logic of the becoming, of the event, of the casual, i.e the logic of a heterogeneous and multiple. "The multiple must be made, not by always adding a higher dimension, but rather in the simplest of ways, by dint of sobriety, with the number of dimensions one already has available – always n-1 (the only way the one belongs to the multiple: always subtracted)".[5] We could say that the rhizome allows a new kind of unity: an under-determinated and lower unity, in an always subtracted dimension to that of its object.

Political implication of the rhizome

This new ontological conception of unity leads us to another dimension of the concept of rhizome: its political implication. In fact, Deleuze and Guattari stresses that the tree has been the image of power, of dominant ontological models, in such fields as linguistics, psychoanalysis, logics, biology, and evolutionary models of classification, all modeled as hierarchical, dualistic or binary systems. As Deleuze and Guattari note, "The notion of unity (unité) appears only when there is a power takeover in the multiplicity by the signifier or a corresponding subjectivation proceeding (…). Unity always operates in an empty dimension supplementary to that of

---

[5] TP: 7.

the system considered (overcoding). The point is that a rhizome or multiplicity never allows itself to be overcoded, never has available a supplementary dimension over and above its number of lines, that is, over and above the multiplicity of numbers attached to those lines. All multiplicities are flat, in the sense that they fill or occupy all of their dimensions: we will therefore speak of a plane of consistency of multiplicities (…). The plan of consistency (grid) is the outside of all multiplicities."[6] So, the concept of the rhizome comes to fight the enemy: the arborescent thought, which is on the basis of centrality, of authority, of state control, and of dominance. The tree dichotomises and produces right ideas and conformed lives.

On the contrary, "the rhizome is reducible neither to the One nor the multiple…it is comprised not of units but of dimensions, or rather directions in motion."[7] The rhizome is neither a system nor a hierarchical structure, nor even a structure made upon units or points. It is rather a plane of heterogeneous forces, of multiple dimensions and in all directions. Rhizome is a mode of organization in which "all individuals are interchangeable, defined only by their state at a given moment - such that the local operations are coordinated and the final, global result synchronized without a central agency."[8] Thus, the rhizome is an alternative system to all authority and hierarchy implied in the tree's image. Being an alternative system, the rhizome puts forward a new order, a new reality. It creates 'lines of flight' therefore offering a political liberation.

The most paradigmatic example of this liberation of the power through the rhizome is Kafka. In their book Kafka – Towards a minor literature, Deleuze and Guattari show that all literature from a minor community is itself a rhizome. We understand that the very first question that Deleuze et Guattari ask is: "How can we enter into Kafka's work?", and we understand also well their answer: "This work is a rhizome, a burrow. (…) We will enter, then, by any point whatsoever".[9] Rhizomatic, Kafka's writing expresses the war machine-book against the State apparatus-book. Kafka and his fantastic bureaucratic machine realize the idea of a book which constructs itself directly as a machinical connection between the writing dispositives and the social machines (juridical and political)[10.]

---

[6] TP: 9-10.
[7] TP: 23.
[8] TP: 19.
[9] Deleuze, G./Guattari, F., *Kafka – Towards a Minor Literature* (K), tr. Dana Polan, Minneapolis/London: University of Minnesota Press, 1986: 3.
[10] The best example of this connection is Kafka's novel *The Castle*. This novel has multiples entrances and, according to Deleuze and Guattari, «the principle of multiples entrances prevents the introduction of the enemy, the Signifier and those attempts to interpret a work

So, once again, rhizome creates a new reality. "To be rhizomorphous is to produce stems and filaments that seem to be roots, or better yet connect with them by penetrating the trunk, but put them to strange new uses. We're tired of trees. We should stop believing in trees, roots, and radicles. They've made us suffer too much. All of arborescent culture is founded on them, from biology to linguistics. Nothing is beautiful or loving or political aside from underground stems and aerial roots, adventitious growths and rhizomes (…). Thought is not arborescent, and the brain is not a rooted or ramified matter".[11] The rhizome constitutes a milieu, a decentred space, with no end or entry point. Deleuze insist: bodies and things ceaselessly take on new dimensions not only through their becoming but also through their contact with different entities.

Thought and diagram

One can say that the tree is an image of thought and the rhizome is the image of a thought without image. As a model, rhizome introduces a new way of conceptualizing thought as becoming, a non-fixed and unruled activity, a contingent reason, a transversal process which is always relation of heterogeneous elements, positive divergence and chaos. Rhizomatic, thought belongs to the "event". Thought happens, and it happens being forced. This is one of the Deleuze's most basilar theses since the first edition of Proust and Signs, in 1964: thought is forced to think and it is not moved by a good will to truth.[12] In this "event" of thought, one must emphasize the rhizomatic way of thinking: not only thought is forced to think chaos, but also it is rhizomatic in itself, it is a composition of chaos.[13] Not only it is

---

that is actually only open to experimentation» (K: 3). For a more developed analysis of Kafka's minor literature, see Pombo Nabais, Catarina, *Gilles Deleuze. Philosophie et Littérature*, Paris: Harmattan (in press).

[11] TP: 17.

[12] In fact, already in *Proust and Signs*, Deleuze stresses the violence which forces thought. As he writes : "The sensuous sign does us violence: it mobilizes the memory, it sets the soul in motion; but the soul in its turn excites thought, transmits to it the constraint of the sensibility, forces it to conceive essence, as the only thing that must be conceived. Thus the faculties enter into a transcendent exercise, in which each confronts and joins its own limit: the sensibility that apprehends the sign; the soul, the memory, that interprets it; the mind that is forced to conceive essence". (Deleuze, G., *Proust and Signs*, tr. Richard Howard, Minneapolis: University of Minnesota Press, 2000: 101).

[13] Deleuze and Guattari define chaos as a virtual which is both birth and dissipation of all possible forms. "Chaos is defined not so much by its disorder as by the infinite speed with which every form taking shape in it vanishes. It is a void that is not a nothingness but a *virtual*, containing all possible particles and drawing out all possible forms, which spring up only to disappear immediately, without consistency or reference, without consequence"

created but also it is a creative process in itself. We could then say that thought is an experimental process, a becoming-thought. Thought is in constant movement and relation to the Outside, to chaos. Thought is confrontation with chaos, but it is also made from chaos. Thinking is the experience of becoming other with chaos.

In this way, we can understand Deleuze's conception of the diagram, not as a fixed, stable and organized model of thought, but precisely as a creative and rhizomatic process. Diagram is the Thought's process of creation itself. In fact, Peirce is a reference for Deleuze. Deleuze recognizes Peirce as the "true inventor of semiotics",[14] but he does not accept his reduction of diagram to icons, indexes and symbols. For two main reasons: because index, icons and symbols should not be seen as signifier-signified relations but as territoriality-deterritorialization relations. "Diagrams must be distinguished from indexes, which are territorial signs, but also from icons, which pertain to reterritorialization, and from symbols, which pertain to relative or negative deterritorialization".[15] Secondly, because diagram has no representative nature at all: "the diagrammatic or abstract machine does not function to represent, even something real, but rather constructs a real that is yet to come, a new type of reality".[16] Diagram is a creative device. Not a graphic representation of relations but a creative agent of reality As Deleuze explains, the diagram acts as a modulator, a conjunction of matter and function. For Deleuze, the diagram is not a static model, but a constructivist performance or experience of becoming.

So, thought as a creative process is a rhizomatic diagram. And instead of Kantian's table of categories from Critique of the Pure Reason, where thought represents the world following strict and static rules, one would have in Deleuze a kind of a Diagrammatology of the rhizome-thought: a thought that happens by chance, through the violence of the problems that force it and constructs itself, modulates itself to its problem. Thought is then an actualization of the solution for each problem it faces by accident. Shortly, a thought that creates reality, both of the world and of itself (as modulating itself to the problems it has to face and that forced it to think at the first place).

But, diagram in Deleuze also refers back to the problem of chaos in thought. "The diagram is indeed a chaos, a catastrophe, but it is also a germ

---

(Deleuze, G. /Guattari, F., *What is Philosophy?* (WP), tr. Graham Burchell and Hugh Tomlinson, London/New York: Verso, 1994: 118).
[14] TP: 586, quote 41. See also Deleuze, G., *Francis Bacon – Logic of Sensation* (FBLS), tr. Daniel W. Smith, London/New York: Continuum, 2005: 109.
[15] TP: 157.
[16] Ibid.

of order or rhythm."[17] Let us take the paradigmatic example of painting, which Deleuze defends to be the strongest experience of catastrophe in all arts. As Deleuze says, the diagrammatism of painting supposes the experience of catastrophe. Because diagram is nonrepresentational and non signifying, painters are able to overcome the usual organization of perception (clichés) and to create a new pictorial order. And that new order is a complex emergence out of chaos, of the elements of rhythm, with its territories and milieus.

The struggle both against and within chaos in Art, Philosophy, and Science is also one of the central themes of What Is Philosophy?, notably in its final chapter, "From Chaos to the Brain". For Deleuze and Guattari, there are three forms of cutting out chaos. Art, Science and Philosophy are the three Chaoids, the three forms of thought and the three forms of creating chaos. On each plane that cuts out chaos it occurs an own reality. Within immanence occurs philosophy, within consistency occurs science and within composition occurs art. Science works with chaos trying to order it, to give it a reference, coordinated space and time. From chaos, Science extracts functions in order to organize the world. Philosophy goes the other way: it goes from things to chaos. The work of Philosophy is to give consistence to chaos, to make it a real entity over a plan of immanence. By its side, Art produces works of art as things, not to organize them but to give them back chaos.

These three ways of thinking are less a struggle against chaos than against the opinion and the clichés, against the rules of a straight thinking, always logic. More than fighting against chaos, Art, for example, makes it sensible. According to Deleuze and Guattari "Art is not chaos but a composition of chaos that yields the vision or sensation, so that it constitutes, as Joyce says, a chaosmos, a composed chaos – neither foreseen nor preconceived. Art transformes chaotic variability into chaoid variety".[18] So, Art is a composition of chaos. It transformes chaos into a chaoid variety, moving it out from its state of chaotic variability.[19] Thought is relation to chaos. Not a relation of exclusion, but, on the contrary, of inclusion. Thought is the result of an operation done to chaos, it is the composition itself of chaos. To think is to give consistence to chaos, to give it an inner reality. Chaos becomes Thought, it acquires a reality as Thought or mental chaosmos.

---

[17] FBLS: 72.
[18] WP: 204.
[19] "And what would *thinking* be if it did not constantly confront chaos?" (WP: 208).

Vitalism

The junction of the three planes of Science, Philosophy and Art, is called "brain". Rhizomatic, the brain does not constitute their unity, rather merely their connection, and their chart. The brain is in a state of over flight, self-over flight, it is co-present within all its determinations, and it travels through them at infinite speed. Deleuze and Guattari show us the diagram of the brain by saying that: "It is not a brain behind the brain but, first of all, a state of survey without distance, at ground level, a self-survey that no chasm, fold, or hiatus escapes. It is a primary, 'true form' as Ruyer defined it: neither a Gestalt nor a perceived form but a form in itself that does not refer to any external point of view (...); it is an absolute consistent form that surveys itself independently of any supplementary dimension, which does not appeal therefore to any transcendence".[20] The brain, both as the creation of concepts and the cut of chaos, is the most subtle dimension of a Nature which contemplates, of an internal feeling as a microbrain or an inorganic life of things. This vitalism essential to all form of existence (brains as well as of rocks or plants) is condensed by Deleuze and Guattari in one expression: the inorganic life of things. "Not every organism has a brain, and not all life is organic, but everywhere there are forces that constitute microbrains, or an inorganic life of things"[21].

For Deleuze, it is not man who thinks, but the brain, the brain being a rhizomatic relation to the world. The brain as microbrain is a brain as subtraction of the human condition (n-1). The brain is a contemplation through becoming, it is a way of joining the world by mixing itself with nature, by entering a zone of indiscernibility with the universe. "It is the brain that thinks and not man - the latter being only a cerebral crystallization. We will speak of the brain as Cézanne spoke of the landscape: man absent from, but completely within the brain. Being a "force" or a "form in itself", thought manages to fly over chaos, to make it sensitive, to cut it out so as to turn it into a chaoid. Deleuze and Guattari defend an inorganic life of thought, the microbrain as pure self-contemplation (of itself) without knowledge. Thus, we can say that instead of faculties, Deleuze and Guattari propose the brain, the microbrain, and defend thought as an inorganic exercise of the brain. The inorganic life is an impersonal, pre- and a-subjective life: microbrains present in all Nature, or, to take once again the expression used by Deleuze, the "collective brain" of the small species like plants and rocks. As a collective matter of Nature, rhizomatic thought is a pan-psychism, a natural and neurological theory of the microbrains.

---

[20] WP: 210.
[21] WP: 213.

Nature and thought as rhizomatic multiplicities imply embodiments of life that do not accord with the prevailing biological idea of life. That is why Deleuze created the new complex concepts of "inorganic life" and of the "vitality" of "machinic assemblage." In fact, the concept of rhizome not only changes our perception of Man and thought, but also of Nature and its idea of life. The rhizome collapses the humanistic point of view positioning Man's thought at the same immanent level as animals and even minerals.[22] Far from being the exclusive essence of the organism, life for Deleuze involves couplings between heterogeneous elements (different molecules from different species, strata, or spheres) and the germinal symbioses of such involvements—life's "creative involution." Inorganic life is life that generates more life through non-reproductive couplings between disparate things. Inorganic life generates life and diversity through symbiotic couplings between unliked things that produce something other than themselves—such as the "wasp-orchid" assemblage or the "animal-virus-human" assemblage: transversal communications which biology views as tangential to evolution. "Not every organism has a brain, and not all life is organic, but everywhere there are forces that constitute microbrains, or an inorganic life of things".[23] So, Deleuze proposes a new paradigm of life: a life that no longer belongs to the pre-existing biological framework, rather to a "creative involution" model. Within this model, there are assemblages of both organic and inorganic life, so we can say that Deleuze's vitalism is an ontology of the inorganic life.

According to this vitalism of a 'creative involution', Deleuze says: "Never send down roots, or plant them, however difficult it may be to avoid reverting to the old procedures (…). What is lacking is a Nomadology, the opposite of a history".[24] Instead of a top-down, vertical thinking that delineates being in higher to lower levels of categorical perfection, thinking becomes horizontal and experimental as Deleuze and Guattari articulate a virtually limitless connectivity between heterogeneous beings. Instead of specific genealogical lineages of origin, selection, reproduction, and evolution, they map a non-teleological and unpredictable network of symbiotic alliances, trans-species affiliations, symbiogenesis, and co-evolution. Instead of a history of the organic (based on the Same and the

---

[22] In a complete radical anti-humanism program, Deleuze and Guattari almost touch the paradox: in the will of destroying all idea of subjectivity, they proclaim the thought-brain itself as a subject. Of course, it is not paradoxical, because the subject they are refereeing to, is an immanent and inorganic subject, with no subjectivity at all, with no transcendence of any kind. Formally paradoxical, the most radical way of deny subjectivity is by defending a immanent and inorganic subject.
[23] WP: 213.
[24] TP: 25.

Subject), Deleuze defends a geophilosophy of the inorganic based on the heterogenous multiplicity. Nomadology is then a geophilosophy which affirms how the earth moves in flows and folds, and how it stratifies and deterritorializes, with a constant and creative instability, that we should discern in human social stratification. Geophilosophy offers more than just a description of matters and forces; it articulates an ontology that maps a rhizome both in nature and the brain and thought as a whole network of matter-energy flow. Geophilosophy is then an ontology of immanence - or a non-linear and complex enfolding of nature and thought.

The principles of the rhizome

We are now able to understand the principles of the rhizome, which at the same time summarize and extend what we have analyzed. Principle of connection (1) and heterogeneity (2). In a rhizome, an unspecified point can be related to all the others. Dichotomy, opposition and arborescent order (where ramification and links are made of contiguous points with equal dimension), all that gives place, in the rhizome, to multiple and heterogeneous chains of connection, without an axis or a central structure. Because it is decentred, the rhizome makes possible the crossing not only of various dimensions but also of heterogeneous points. Deleuze takes the example of linguistics and stresses that the principle of connection means that language, within the model of the rhizome, is also connected to non-signifier elements as political, social and bureaucratic elements.

Principle of multiplicity (3). Multiplicity is the multiple when treated as substantive. It has nothing to do with the divisions, pseudo-arborescent multiplicities, frequently mentioned as composing the One or as dividing the subject. Deleuze sketches the formula: Tree=One or $N+1$ which means that the tree model is a totality itself made of units. It is $N+1$ because the organizing model and meaning principle means transcendence. N+transcendent. On the contrary, Rhizome=$N-1$, it is like a substraction because it functions above and under all forms of subjectivity. It is a Multiplicity to which we have already withdrawn the transcendent principle. Rhizome is a Multiplicity with N dimensions, always in nomadic movement. It implies the absence of a central unity, either that of the subject, or that of the object. In fact, multiplicities have neither subject nor object. They change their own nature on the basis of the rhizomatic connections they establish with other multiplicities. They are defined by connections with the outside, the plan of immanence, deterritorialisation and lines of flight.

Rhizome is then a process not of differentiation but of symbiosis, of connection of heterogeneous elements.

Principle of a-meaning rupture (4). Contrary to a tree, which has final and meaning cuts, a rhizome does not have marked and meaning ruptures, which separate segments or structures. In a rhizome, what is blocked, broken and interrupted, takes its connections again through others lines, in particular those of deterritorialisation or flight lines. However, rhizome also has lines of segmentarism which stratify, organize and give it a direction. Even its lines of flight, which do not stop establishing connections, sometimes lead it to new meaning formations that re-stratify and reconstitute the subject. These constant movements of deterritorialisation and reterritorialisation follow an a-parallel evolution by making the connection of heterogeneous elements and different degrees of differentiation possible. The evolution is made not by descent, copy, imitation or identification (as in the case of the tree model), but by becoming, i.e. contagion, variation, capture, expansion. To become is to let molecules of getting into a neighbourhood, is to connect beings of different scales and various species. To become is not to progress, to evolve or to regress. To become is not a question of past or future, of clean dimensions of molar entities from the plan of transcendence. To become is always to become unlimited-present and to become-minority[25] in a plan of immanence (N= -1).

To become or to involute is to be according to lines of flexible segmentary nature, lines of flight. All becoming is already molecular, involutive movement in the plan of consistency, oscillation or even overcoming of the binary machine, that is, the molar organization. As Deleuze and Guattari say in A Thousand Plateaus, "starting from the forms one has, the subject one is, the organs one has, or the functions one fulfills, becoming is to extract particles between which one establishes the relations of movement and rest, speed and slowness that are closest to what one is becoming, and through which one becomes".[26] Becoming is not the

---

[25] As Deleuze explains, majority and minority don't mean quantity. They rather refer to sets, states, owners. The majority, domination state, is made up from the organizing movement, from the central point, from the stating subject: "Following the law of arborescence, it is this central Point that moves across all of space or the entire screen, and at every turn nourishes a certain distinctive opposition, depending on which faciality trait is retained: male-(female), adult-(child), white-(black, yellow or red); rational-(animal)" (TP: 323). So, minority cannot be confound with minority as a state, as a quantity of people smaller than the majority. As Deleuze says in *Negotiations*: "What defines the majority is a model you have to conform to: the average European adult male city-dweller, for example... A minority, on the other hand, has no model, it's a becoming, a process." (Deleuze, G., *Negotiations, 1972-1990* (N), tr. M Joughin, New York: Columbia University Press, 1995: 173).
[26] TP: 300.

production of something else than itself. "Becoming is a rhizome, not a classificatory or genealogical tree".[27]

Principle of cartography (5) and calcomanie (6). Rhizome is not about copy or reproduction. These proceedings belong to the representative model of an evolutionary binary process, based on a genetic axis which articulates and differentiates the copies within a surcodifing structure. Copy stratifies, organizes and blocks multiplicities according to axes of significance and of subjectivation and, for that, "when it thinks it is reproducing something else it is in fact only reproducing itself. That is why the tracing is so dangerous. It injects redundancies and propagates them".[28]

The rhizome is a map, an opened chart, suitable for connection, which does not take the world as a reality to be reproduced, which does not copy reality but which builds it through experimentation. In fact, map is a matter of experimentation and performance, and that is what distinguishes it from tracing, which is a matter of static knowledge and competence. Map is a process, is a modulation within reality. On the contrary, tracing is a stabilization of the reality into an image. By building the real, always within immanence, rhizome also builts itself. "The map is open and connectable in all of its dimensions; it is detachable, reversible, susceptible to constant modification. It can be torn, reversed, adapted to any kind of mounting, reworked by an individual, group, or social formation. It can be drawn on a wall, conceived of as a work of art, constructed as a political action or as a meditation".[29] The map does not have language or sign as representative authority nor the subject-author as stating and expression structure.

***

As a way of concluding, we may say that Deleuze not only understands Nature as a rhizome, as a non-organized world, but also takes this image of Nature and applies it in the thought. He does make an animistic operation: he de-subjectives human thought by proliferating Nature in it. He does extend thought to the immanence plane of Nature and that is why he says that there are micro-brains everywhere in Nature: from humans to minerals and flowers. Thought is then an animistic event, an immanent rhizome that proliferates in both Nature and Man. And vitalism is a dynamic diagrammatology of this new category: the rhizome-thought.

Rhizome is then the image-concept of the ontology of Nature which also includes thought. The anti-humanist program is here accomplished in its

---

[27] TP: 263.
[28] TP: 15.
[29] TP: 13-4.

most extreme formulation. It becomes a cosmological program, a study of the inhuman forces, a topology of the inorganic life from rocks and plants until the nonhuman becoming of man. We could say that the rhizome as the diagram of this new image of Thought, a thought without image, is a kind of an immanent diagrammatology that crosses Nature as well as Man, or that takes human thought as an expression of a Nature's activity. So, the concept of rhizome is a whole new diagram of thought, nature and politics. It threatens the arborescent western culture and proposes new ways of thinking biology and life itself, it changes the concept of thought and chaos, and by its multiple and subterranean proliferations it is an alternative to political power.

Deleuze and Guattari want to kill the tree in our head and clear the way for an alternative concept that cultivates rhizomatic complexity. Instead of the icon of the Same, he urges that we perceive the multiplicity of differences: the different bodies and different assemblages (aboriginal and non-aboriginal, social, cultural, political, economic, spiritual, and aesthetic). I would like to finish with a magnificient quote from A Thousand Plateaus: "Many people have a tree growing in their heads, but the brain itself is much more a grass than a tree".[30]

References

Deleuze, Gilles, *Francis Bacon – Logic of Sensation*, tr. Daniel W. Smith, London/New York: Continuum, 2005.

—— *Proust and Signs*, tr. Richard Howard, Minneapolis: University of Minnesota Press, 2000.

—— *Negotiations: 1972-1990*, tr. M Joughin, New York: Columbia University, 1995.

Deleuze, Gilles and Guattari, Félix, *A Thousand Plateaus*, tr. Brian Massumi, London/New York: Continuum, 2004.

—— *Kafka – Towards a Minor Literature*, tr. Dana Polan, Minneapolis/London: University of Minnesota Press, 1986.

—— *What is Philosophy?*, tr. Graham Burchell and Hugh Tomlinson, London/New York: Verso, 1994.

---

[30] TP : 17.

Pombo Nabais, Catarina, Gilles Deleuze. *Philosophie et Littérature*, Paris: Harmattan (in press).

Oxford English Dictionary. ed. J.A. Simpson and E.S.C. Weiner. 2nd ed., Oxford: Clarendon Press, 1989, OED Online Oxford University Press.

www.ingramcontent.com/pod-product-compliance
Lightning Source LLC
Chambersburg PA
CBHW071434150426
43191CB00008B/1121